LINER DMV WILLEM RUYS

INTRODUCTION

The Stoomboot - Rederij "Rotterdamsche Lloyd (steam shipping company "Rotterdam Lloyd") was founded in January 1875. Managed by Wm Ruys and I.G.N. Hoven. The company changed its name in 1883 into Naamloze Vennootschap "Rotterdamse Lloyd" (RL) Ltd "Rotterdamse Lloyd". From 1884 the company cooperated with the Stoomvaart Maatschappij Nederland "Nederland"(SMN) (Steam shipping Company Nederland) which had been founded in Amsterdam in 1870.

First ship of the Rotterdamsche Lloyd: sv Groningen. ***Built in 1874, taken into service 1875, wrecked 1876.*** *(Ill: G.J. Frans Naerebout)*

From 1870 a steamship sailed for the Netherlands Indies on a weekly schedule either from Rotterdam or Amsterdam. This cooperation would later develop into Nedlloyd Lines. The mail contract with the Dutch government in the late 19th century stipulated that both companies had to sail with fast ships.
Just before the turn of the century RL and SMN divided their fleets of liners in a freight service and one especially for passengers.

1924
Slamat

1926
Indrapoera

1928
Sibajak

1930
Baloeran

ss Slamat *(1924-1941; 11.406 gross). Docked in 1931 and fitted with a new 42 foot bow. Served as a troop transport in WW2. Was in 1941 attacked near Crete by German planes and set on fire. 193 men died in this attack.*

ss Indrapoera *(1926-1956; 10.825 gross). In 1932 lengthened at the Schelde dockyard and equipped with diesel engines. Her beautiful interiors were destroyed in 1940 when she became a troop transport. In 1949 converted into a freight/passenger ship.*

At the end of the 1930ties the Rotterdam shipping company Rotterdamse Lloyd had five large passenger liners at their disposal exclusively for the Dutch Indies lines. These five had a combined capacity of 18.744 passengers. However annually only about 12.500 passengers were transported. So in fact there was no necessity for a new ship. However the company foresaw a fleet renovation by 1941. They were urged on by the initiatives of other Dutch shipping lines. The Holland-Amerika Lijn had left the economic crisis behind with the majestic ss *Nieuw Amsterdam* followed by mv *Oranje* of Stoomboot Maatschappij "Nederland" from Amsterdam.

ss Sibajak *(1928-1959; 12.040 gross). In Allied service in 1941. The interior converted for troop transport use in Singapore. Back under the company flag converted into emigrant ship.*

On 16 July 1938 the Rotterdam Lloyd launched its plans to build a large passenger liner. The engineering department expressed this aspiration through the "Begrooting 3086a". 'Estimate 3086a'.
On the 12th of September of that same year the commission to build the ship was granted to the yard that up to then had built nearly all passenger liners for the shipping company. Royal dockyard "De Schelde" in Flushing. On 25 January 1939 the keel was laid under number 214.

Sister ships mv Baloeran *(1930-1943; 17.001 gross) and mv* Dempo *were RL's last new great passenger ships before the outbreak of the Second World War. Both ships would be destroyed during the war.* ***Baloeran*** *hit a mine after she had been commandeered by the occupying forces and converted to a hospital ship. She was grounded 1.3 miles from the Northern Pier at IJmuiden and destroyed by British planes. Remains of the wreck are still there.*

(Drawing: D. Borchert)

The designs for Baloeran / Dempo *are exactly ten years earlier than those of Construction number 214.*
Both ships could be distinguished as ships from the 1920ties. They could among other things be recognized by the straight stem and cruiser stern.

Mv Dempo *(1931-1944 16979 gross) She was converted for the transport of 2000 soldiers immediately at the outbreak of World War II. Deployed at the landings in Italy. Torpedoed on March 17 1944 on the coast of North Africa near Algiers. All those on board were safely put ashore.*

WAR

The Netherlands prepared for a possible war. Rotterdam Lloyd had an air raid shelter built on its quays, an air defence service was organized and an armament program for the new passenger ship under construction then known as `Bouwnummer 214' was designed.

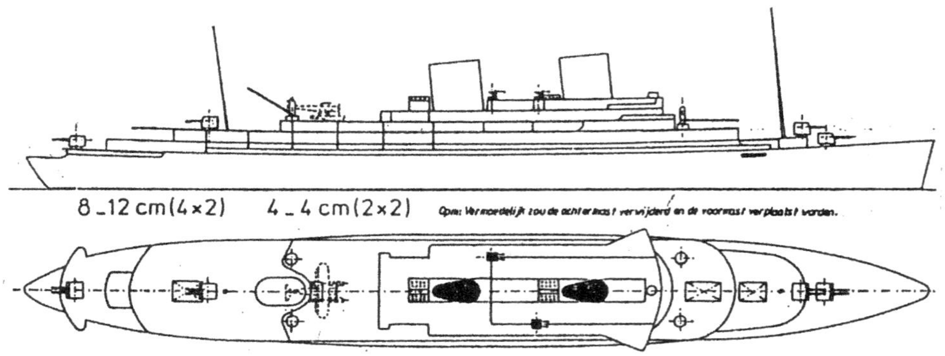

Design `KMS214` as auxiliary cruiser. The rear mast would probably have to be removed and the front mast moved.

On May 17 1940 the German occupier confiscated the `Schelde dockyard. 214 could be launched but the yard management convinced the occupying forces to keep the hull on the stocks. 214 would immediately become a target for bombers if she was in the water. The dockers thought of ways to prolong or slow down further construction. Four out of the eight Sulzer engines that had been ordered disappeared into the countryside as well as many parts and materials from suppliers. Or they were stolen from under the noses of the German plunderers. The resistance sabotaged countless attempts by the enemy to destroy the ship. The Dutch government in London urged the allied forces to be careful during bombardments. *214* escaped many direct hits and in November 1944 when Flushing perished in a sea of fire *214* lay apparently unscathed in the middle of an indescribable sea of destruction. The yard suffered millions of guilders of damage. No building had its roof intact, there was no glass in the windows, not a single crane could be moved.

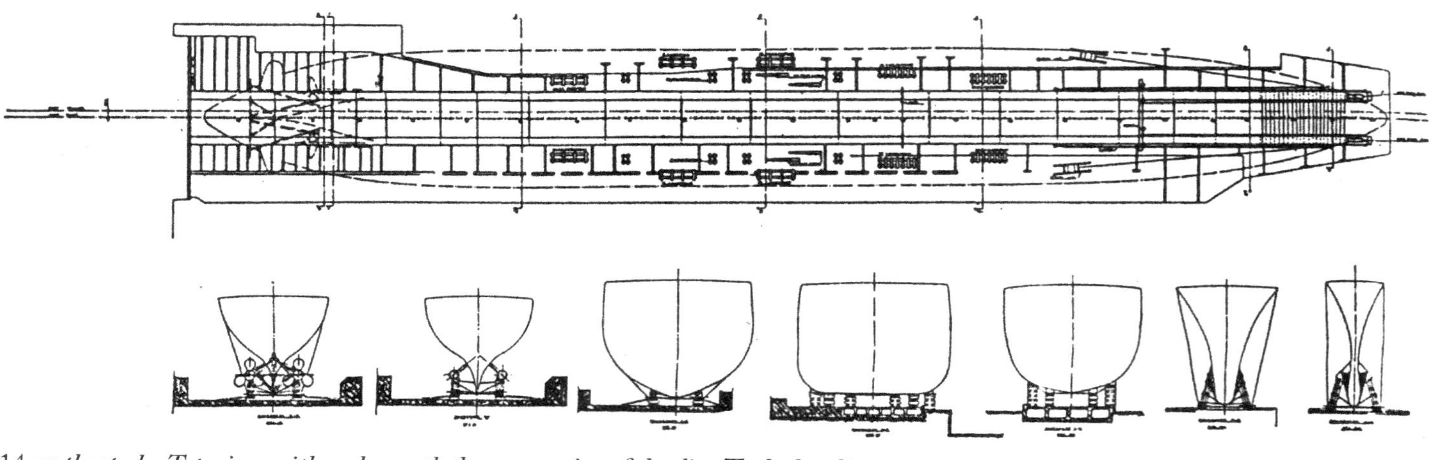

214 *on the stocks. Top view with underneath the cross section of the slip. The hydraulic rams.*

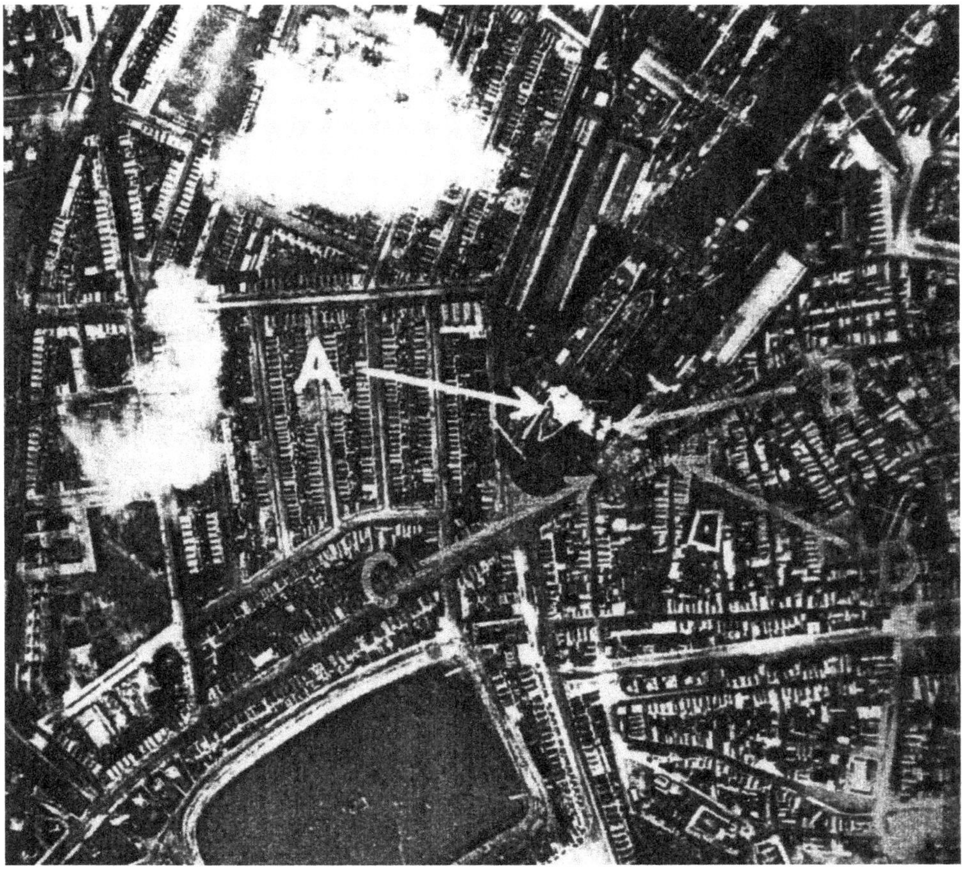

Proverbial luck shown. On the slipways 214 (see the arrow at A) and mv Borneo *(under construction for SMN) the last one suffers a direct hit.* 214 *is unscathed.*

After the liberation in November 1944 when Flushing was without fresh water the sea water evaporators of *214* worked at full capacity, 300 cubic meters a day. 214 was the symbol of hope for better days for the city of Flushing and in fact for the whole of the province of Zeeland. Rotterdam Lloyd lost three of its passenger ships during the war. Besides that *Indrapoera* could only be used again after extensive repairs and it would take years before *Sibajak* would return to the fleet. *214* awaited a busy life.

Map of the "De Schelde"complex in Flushing 1947. The North slip Noordhelling is shown horizontally to the left from the Dock. The circles show the flight of the cranes. Sheds on both sides of the slips, welding sheds and moulding lofts next to them to the right.

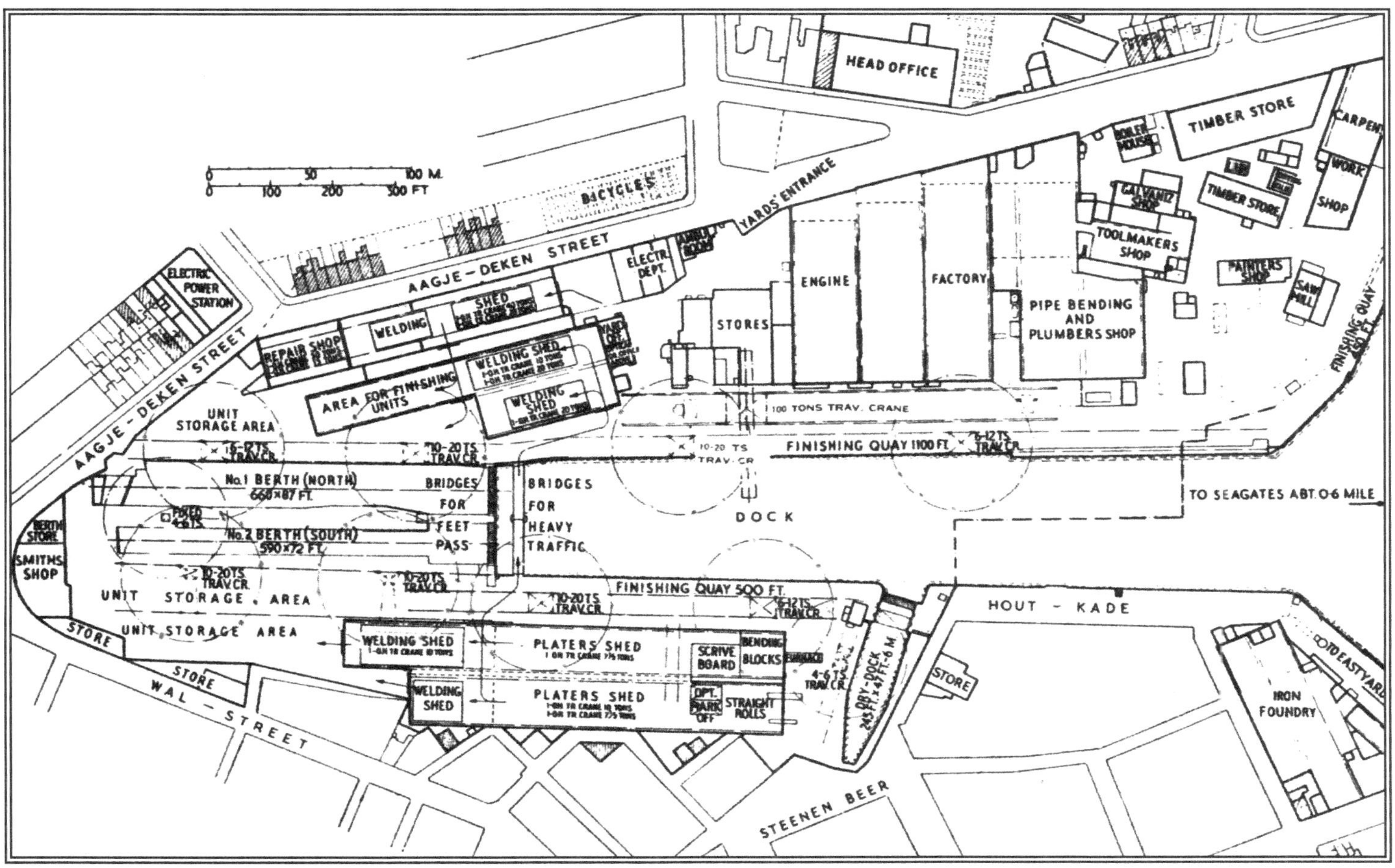

CONSTRUCTION

Shipyard 'De Schelde' in Flushing

Shipyard "De Schelde" had been the main supplier of passenger ships for the Rotterdam Lloyd since 1884. One of the few companies in the Netherlands that could build a complete passenger liner. The urge for bigger and faster in shipping carried consequences for the builders as well. When the order to build *214* had become a fact in September 1938 the yard adapted its workshops, cranes and other facilities where needed and extended them if necessary. The 'Noordhelling' was too short and had to be extended including a tunnel so that traffic to and from the workshops could continue without any hindrance. Two tower cranes had to be removed for the wide ship. The slipway sill and the slipway lock doors on the harbour side and the connecting bridge were widened by some meters. An electric elevator was placed beside the high stem so that the yard workers could reach the higher decks without getting exhausted.

After the liberation the yard was found to be in a deplorable state. A year later most of the damage had been repaired. The yard offered work for 4000 employees and the machine building facility for steam engines, diesel engines and turbines performed to the best of its ability. The yard workers had managed to keep a lot of material and parts hidden from the Germans so that finishing *214* could start.
The drawing room managed by eng. C. Kapsenberg had finalized the plans. A 1000 workers and an additional 100 draughtsmen and constructors would devote six million hours of work to that.

Bottom section with automatically electrically welded tank top.

A riveter team in the workshop. From left to right.: the heater, the holder-up and above the riveter with his gun.

Immediately after the keel was laid on 25 January 1938 the double bottom sections were placed and connected. The welded and partially riveted sections of the keel weighed 24 tonnes. These were followed by the bottom and the sections that had been welded in the workshops. Every section consisted of three floors half the width of the ship including the deck plating. In total 32 of these sections were constructed on either side of the vertical keel.

Numbered rivets, marked with white paint.

The complete tank top and the engine room were welded. The butt ends of the bottom. The double bottom, the plates of the hull and the decks as well as the oil bunkers, all foundations of the auxiliary engines, the stabilizer tank including the internal stiffeners. The welding technique meant that the superstructure could remain light-weight. Three million electrodes were used in the built, still no less than 1.400.000 rivets played a role of importance. Up to and including the A-deck the overlaps have been riveted and the butts welded. So, the hull of *214* is a combination of riveting and welding up to and including the A-deck.

Welder. Inside the welding apparatus an electric current generates heat. This welds or melts the plates together.

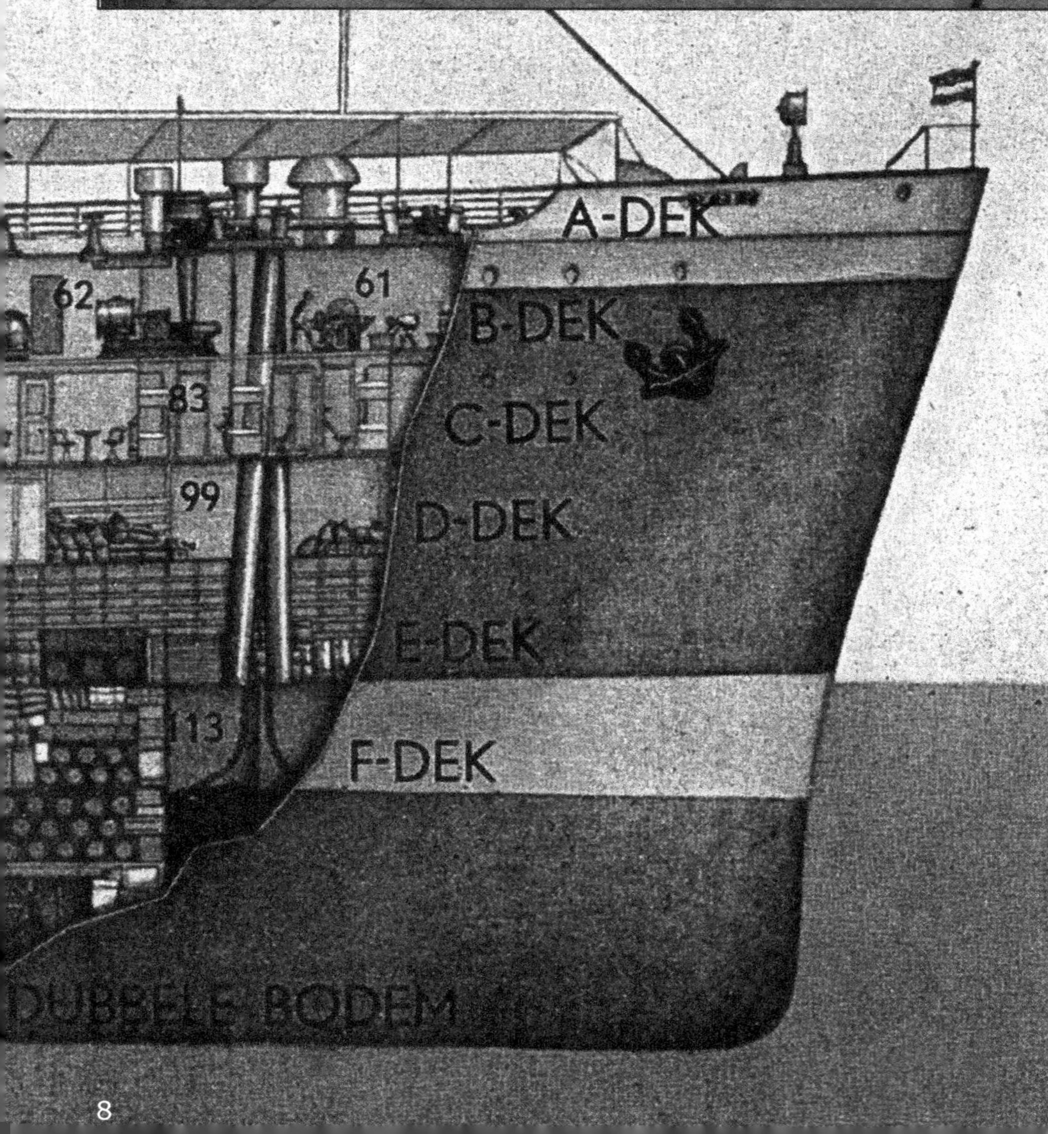

On the template floor points are marked of one of the cross sections of the ship. These points are connected by wooden strips. If all the projected points are correct and there are no false bends, in other words, 'the line squares' then this is scratched into the plate along the strip.

The draughtsman draws technical drawings that are transferred to the actual size in the moulding loft. These are made into wooden templates that serve as the model for the cutting of the steel plates and casting of iron or copper alloys. The plates are bent into the right shape with hydraulic presses: these have been returned in such a way that there is very little waste. On the slipway the bottom plates with ribs are laid on wooden keel blocks. Then gradually the ribs rise on which the plates are hung. Wooden supports prevent the hull from falling over. In the shed the plates to be used lie on long frames. A punch makes the holes for the rivets. To make it known to the puncher where he should punch the holes the plates have been scored. The spots indicated with a chalked mark. It is all about accuracy. The puncher should not be in a hurry. In spite of this handicap an experienced puncher may punch between 3000 and 3500 holes a day.

Moulding. In the moulding loft the ribs and other parts are drawn in their actual size on the floor after which templates are made.

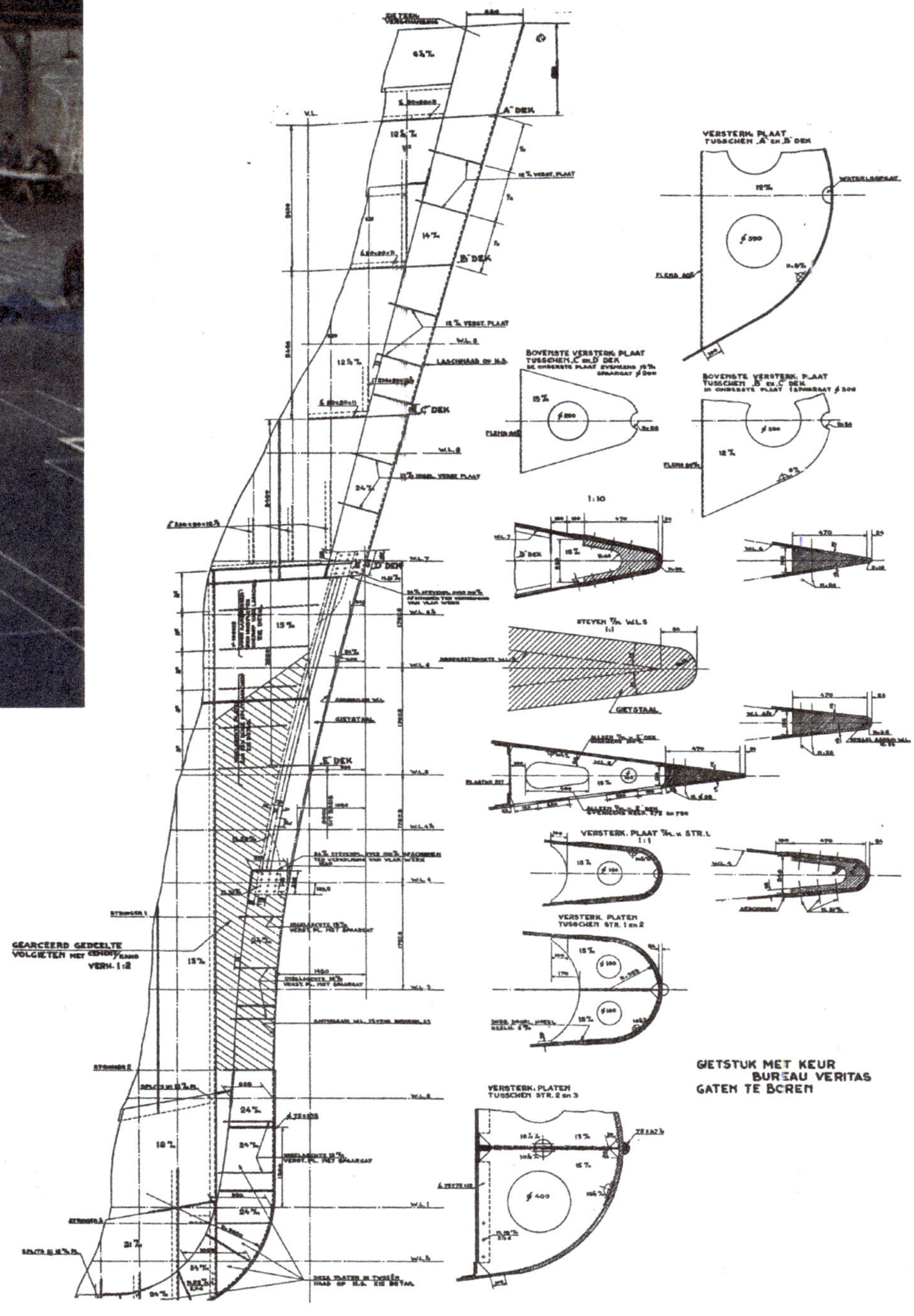

The cutter cuts away along the chalked lines the superfluous rims with a heavy cutting tool or instead the blowtorch is used to burn away any material that is left. Any manholes in the plate are cut out in one movement by a heavy press.

Tremendous forces are at work on the hull of a ship. Every ship must be capable of sustaining those forces. Every shipbuilder can calculate in advance what the forces are that the ship designed by him can endure. He can burden the ship with a theoretical load and calculate what the bending stresses are as a result. The problem with a bow is that it also has the function of an attack on the elements, especially water. The stem should facilitate fast progress through the water. In the case of *214* the choice was made for a raked stem with a modest bulb. This invention had been used earlier on the French passenger liner *Normandie* (1935). Through the swelling at the foot of the stem resistance is diminished by a flattening of the wave patterns. In later years the simple shape that had been applied on *214* would become a large size cigar shaped bulb. On these pages the idea of the *214* bulb stem and what it involves is shown.

PENOLA BASEKOTE 31 · PENOLA SLIPKOTE 4

Op deze minerale hellingvetten zijn in Nederland het m.s. „WILLEM RUYS" en nadien vrijwel alle grotere schepen en vaartuigen te water gelaten.

Standard Amerikaansche Petroleum Compagnie - „Esso-gebouw" - 's-Gravenhage

ms „Willem Ruys"

The rudder is placed on the stern. It rests in the pintles of the straight part, the sternpost. The stern is vulnerable especially with incoming waves or swell. It is the rudder that is put under pressure. The rudder is installed after the propellers. The rudder turns on the mounted bearings. These are the bolts that connect the rudder with the stern. At the bottom it rests in a pivot. Rudders are made of the best quality material as is anything on board a ship. Quality comes first. The cast iron rudder frame of *214* was made, according to the text with this picture, in Belgium at the foundry of Usines & Acières Allard at Mont sur Marchienne. Many hundreds of companies were involved with the construction of a giant like this future flagship for a prestigious shipping company like the Rotterdam Lloyd and therefore besides the yard and ship's staff many thousands of other wage earners.

The hull is painted in the Company colours while still on the slipway.

The launch is set in motion by a woman's hand who 'removes 'the last impediments'. These are a number of levers, the last one called the 'latch', a sturdy piece of iron that secures the ship till the last moment. The man who operates the latch is seen here sitting under the vessel's 600.000 kilograms of steel.
Heavy pounding on the ship tells him that the bilge blocks are removed and the ship sits on the greased channels. A whistle sounds and then lots of things are set in motion because the man under the ship sees to it by a stroke of the riveting hammer that the tight rope along the keel is loosened after which the heavy object on top of him sets itself into motion. Faster and faster. After some seconds there is daylight over the man. The rising stem glides past him and some seconds later 214 just baptized as *Willem Ruys* is in the water. 1 July 1946.

It goes without saying that finishing the complicated passenger liner was no mean feat considering the difficult circumstances immediately after the second world war with its shortages of raw materials and other products. One had started on the interior in dribs and drabs while the ship was still on the slipway but as soon as she was on the quay outfitting went crescendo. The most difficult part was installing the engines. These were built and tested in the factory beforehand. Four of the main engines of mv *Willem Ruys* were present. The other four that had disappeared during the war to Gorinchem arrived one month after the launch. When the new ship had to leave Flushing she was not completely finished.

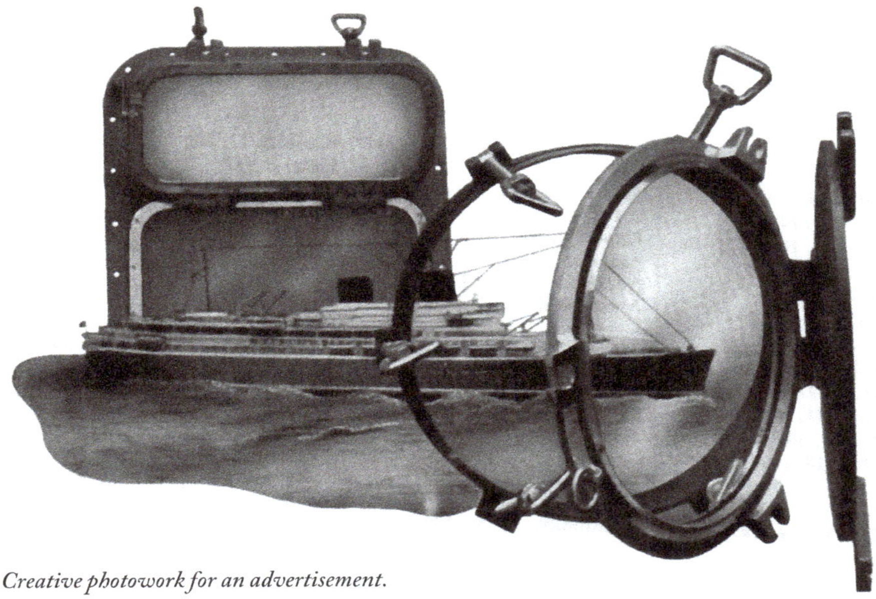

Creative photowork for an advertisement.

Below: Ready for launching. *(drawing: D. Borchert)*

mv Willem Ruys *at the outfitting quay of 'De Schelde' in Flushing. The different work sheds of private enterprises or subcontractors who are involved in the outfitting and finishing. There are cranes for heavy and light loads. Along the quay provisions are made for electric cables for light and power, mains for pressured air, gas and oxygen, water and steam.*

Plans for modelling

The Dutch Modellers Association offers plans for constructing a model. See www.modelbouwers.nl

❶	10.10.002:	Passengership Willem Ruys 1/200 3 sheets
❷	10.10.006:	Waterlinemodel Willem Ruys 1/500
❸	16.10.004:	Passengership Willem Ruys 1/100 Shipyard plans, 6 sheets A0

Bouwnummer *214* had taken to the water without hesitation as *Willem Ruys*. Stern first as it should. Why are ships launched stern first? Because, at the moment that the ship has entered the water halfway the front is subjected to a lot of upward pressure. The weaker stern with vulnerable parts such as the propellers, the rudder and the propeller frame could easily be damaged. Stern first through the slipway lock doors proved to be quite an effort. The passage is 25 metres, *Willem Ruys* measures 23.34 metres. Five tugs reserved two days for the job on 28 September 1947 but they finished the work in less than seven hours. *Willem Ruys* is seen here passing the locks in Flushing. Three tugs at the stern and two at the stem. (not in view.) (Picture J.M.Matena.)

To sea for the first time. The people of Zeeland showed up in large numbers to watch her leave. The ship had been in their midst for nearly eight years. Their *Willem Ruys* set of to roam the world and never to return.

During the trails(15-18 November 1947) it appeared that the prolonged stay on the slipway and the violence during the war had left their mark on the construction. Once at sea the ship did not fully perform as to the expectations of the classification societies and the shipping company. There also arose disagreements concerning payments which dampened the relation between the builder and the shipping company. The new ship achieved a maximum speed of 24.6 knots to which 5% could be added in view of her draught in relation to the waters where the speed trails were carried out. Sufficient for a service speed of 22 knots.

Above:
Willem Ruys *outshone in its appearance fellow Dutch-Indies liner* Oranje. Willem Ruys *became with fervour the one but largest but also the merchant ship with most largesse of Rotterdam and the Netherlands.*

Left:
21 November 1947. Changing the colours. The solemn moment of conveyance from wharf to client.

W. Ruys B.Ezn (1894-1942).

NAME

The passenger liners of the Rotterdam Lloyd were usually named after volcanoes in the Netherlands Indies. The management of the company hesitated between *Batoer* and *Ardjoeno* as the name to be for their new flagship. The German invasion meant that the decision was postponed. The future vessel was indicated as *P6*, later *P7* according to the 1938 plans. As soon as construction started the hull became *Bouwnummer 214*.

The war left a deep mark on the Dutch people and this also showed in the name of this imposing new Netherlands-Indies liner. Instead of following the volcano-tradition 'number 24' was baptized *Willem Ruys*, the name of the general manager of the shipping line who had been murdered by the Germans on 15 August 1942. W. Ruys had denied German officers access to the elite Rotterdam rowing and sailing club 'De Maas'. He was later taken hostage. The management of the Rotterdam Lloyd by using this name also paid homage to the other members of the Ruys family who were named Willem. They had made the company into what it had become. Especially the founder Willem Ruys J.Dzn (1809-1897). He had founded the company that later led to N.V. Rotterdamsche Lloyd. Mv *Willem Ruys* was baptized by the widow E.E.Ruys-van Houten. Mv *Willem Ruys* became a byword. She was the last real great Netherlands-Indies liner. A memory of a proud history.

dss Sindoro *(1900-1932 5471 gross) The first two propeller ship of the Rotterdam Lloyd .Served as a transport for prisoners of war during World War I. (drawing: A Zuidhoek)*

DESIGN

In 1900 Rotterdam Lloyd's dss *Sindoro* followed the developments in the design of the international passenger liner. Accommodation for first class passengers moved from the stern to the quieter amidships. This was made possible by the use of lighter materials that lifted the bridge one deck higher and later again one further up. The higher superstructure offered room for promenade decks in front of the cabins with portholes and windows that could be left open. This profile of the passenger ship was developed further as time progressed.

After 1935 low funnels for motor ships became unfashionable. *Willem Ruys* got tall funnels, two, because mv *Oranje* had only one. (Only a part of the second funnel was in use) The demand for sundecks meant that there was no room for the lifeboats. These were moved down, a striking difference with other ships. Still this is a distorted view, in fact the lifeboats were fitted at the normal height and the superstructure was lifted. The whole superstructure offered a spectacular series of decks. From the A deck up to the top deck with rounded promenades and a wide winged command bridge deck.

Schematic image of aesthetic corrections in the design of a ship's profile.

***Below:** In March 1955, the ship encountered a severe storm in the Gulf of Biscaye. (gouache: A. Zuidhoek)*

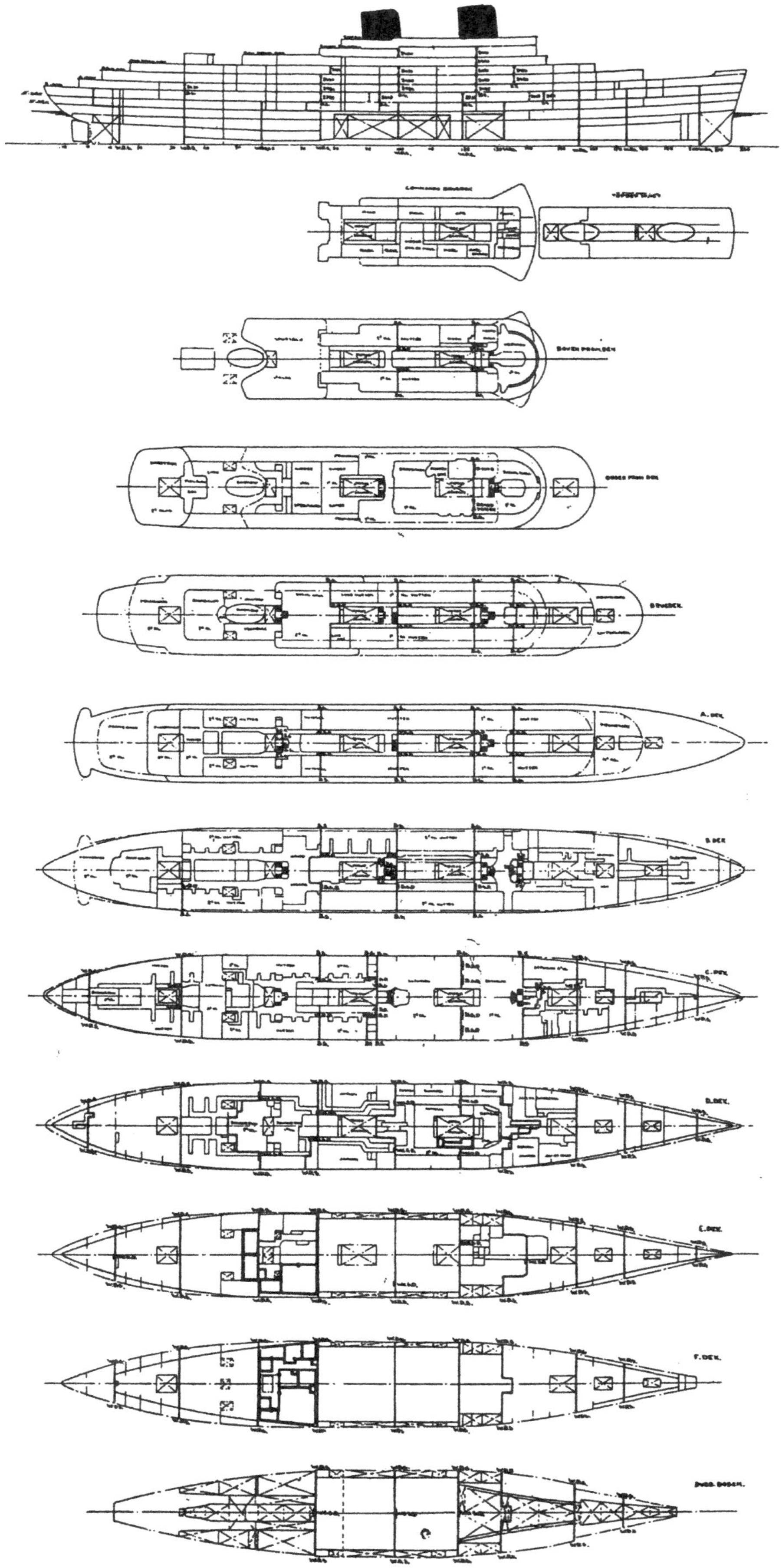

The rounded saloons up to the front bulkhead, the terrace shaped decks, the beautifully shaped funnels and bridge wings, the arrangement of the lifeboats. Those were the defining aspects of Mv *Willem Ruys*. A charming detail were the two extending wings over the narrow spoon stern. The English press called it a 'notable design'.

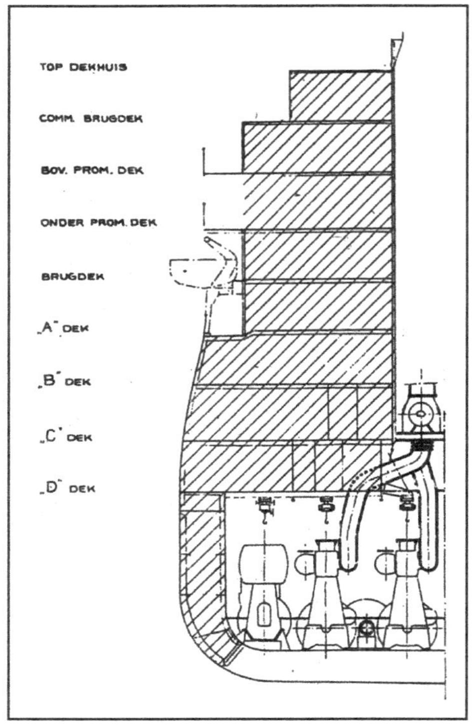

Cross section of the engine room.

Position of fire bulkheads, watertight bulkheads, fire doors and watertight doors in the longitudinal section. Below the deck plans, from top to bottom: command bridge deck/top deck, upper promenade deck, lower promenade deck, bridge deck, A-deck, B-deck, C-deck, D-deck, E-deck, F-deck. Double bottom.

History has shown that Dutch shipping companies well understood how to keep the travelling public interested. The exterior and interior of the ships that plied the routes to the East played an important part in this. *Willem Ruys* bridged the pre-war set with a new era. The plans for the liner were made before World War II in 1939 but changes made during the war years were included. *Willem Ruys* was a combined effort of the construction department of the Rotterdam Lloyd and the engineers at the yard. Ir L.W.Bast was the responsible naval architect.

Willem Ruys with four other liners of the Indies route. At the top the British *Dominion Monarch* (1939 Shaw Savill Line) gross 26.463 and for a long time the world's largest motor liner. She had an enormous freight capacity and could accommodate 508 first class passengers. Characteristic is the well (for loading and unloading through hatch III) between the bridge and the passenger accommodations.

Then mv *Ruys* (1937 KPM/KJCPL. passengers 104. I 331. II 326 III 329 IV gross 14.271) She had remarkably modern lines. She would later sail in the service between the East Indies and South Africa.

Below *Willem Ruys* the lines of a French East Indies liner. The history of its construction is comparable to that of *Willem Ruys*. She was on the slipway when the Germans invaded France. The workers on the yard managed to sabotage the launch until 1944. She was finally delivered in 1949 with the liberating name *La Marsailaisse*. (Messageries Maritimes. 17.280gross accommodation 341 I 75 II).

Bottom. A portrait of dss *Chusan* (1950. P&O-Orient Lines 24.215 gross). Accommodation for 464 passengers first class and 541 second class. This ship could also carry a considerable amount of freight. She followed the same route as *Willem Ruys*, but would head north towards Japan after Singapore. vice versa.

(pencil and watercolour A. Zuidhoek)

ENGINE

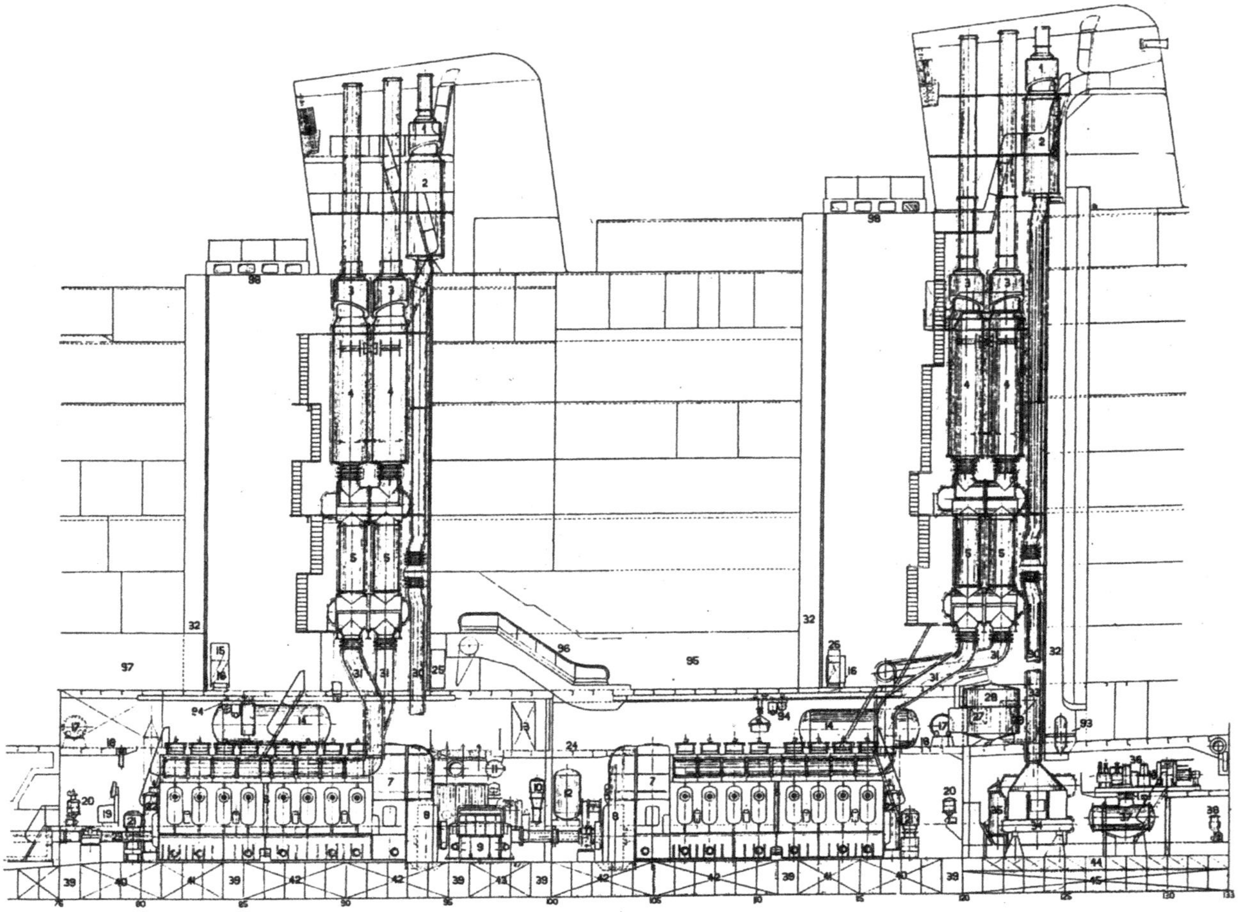

What choice would be made for the propulsion: steam engine, (turbine) or diesel? The choice fell on the latter one. A diesel engine takes up less space than a steam engine. The same choice that was made for Oranje. A Sulzer engine was preferred. A Swiss product. *Willem Ruys'* Sulzer became a technological novelty. Eight times eight cylinderblocks (fast diesels) in two engine rooms. These powered the two propeller shafts through mesh gear boxes. All in all 8x 4000 horsepower. Consuming 5.500 kg fuel per hour. The service speed of 22 knots shortened the distance Marseille Batavia (Djakarta) by two days from 20 to 18. An engine could be stopped during the journey, which was quite an economy on fuel consumption. The horizontal connections between the skin plates were welded instead of riveted to ease progress through the water and the stem was fitted with a bulb.

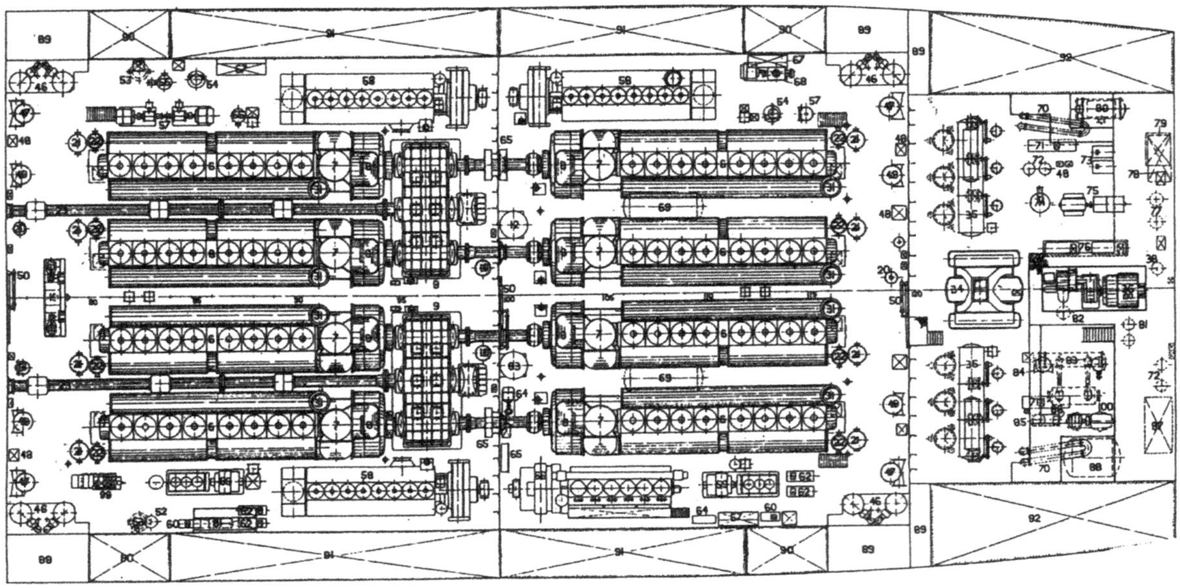

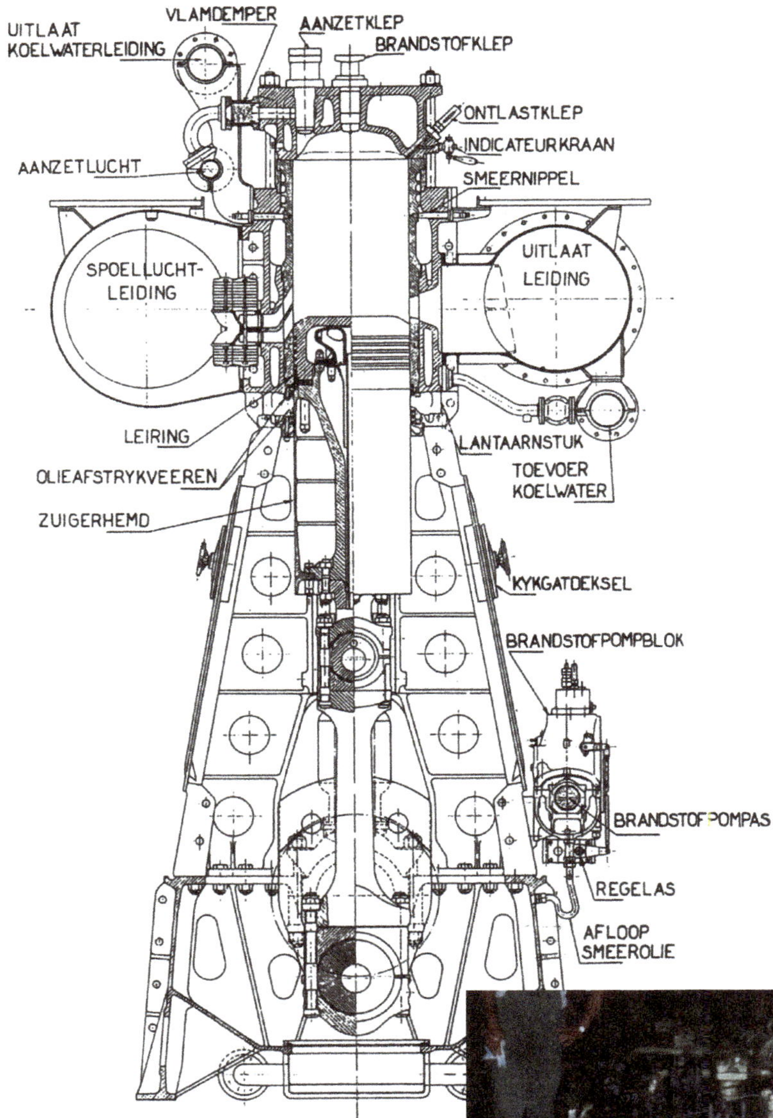

Cross section of one of the main engines. Each cylinder has been constructed as a square block with flanges along the rim. The flanges connect the blocks with bolts so that you have a long cylinder beam which is attached to the frame that by itself is also a stiff compartment. This makes for a sturdy engine that does not have the tendency to vibrate or move.

Cylinders	8
Bore	580 mm
Stroke	840 mm
Horsepower	4000
Revolutions	215 per minute
Pressure	4.75 kg/cm^2
Piston speed	6.0 m/sec

Ir D.T. Ruys was responsible for the machine engineering part of Mv Willem Ruys. Thanks to his design the height of the engine rooms could be kept that low that two decks over the engine rooms could be extended . The space gained was used to house the galleys. *(drawing. A. Zuidhoek.)*

Six of the engines have been made by ‘De Schelde’ in Flushing licensed by Sulzer. Two were supplied by Sulzer in Winthertur. The engines were fitted with the renowned Sulzer flush ports at the bottom of the cylinders.

Both engine rooms take up two decks vertically from the middle of the ship. The two departments of the main engine room reached up to E-Deck. These have the main engines, the eight two-stroke Schelde-Sulzer diesels. This room also has three auxiliary engines with 8 cylinders each and one of 6 cylinders coupled with a dynamo needed for the ship’s electricity. The dimensions of the engine rooms, especially the small height of the engine room installations is demonstrated in the drawings beside and on page 18.

View of the top of the cylinders in one of the blocks. *(picture A. Zuidhoek)*

It had to be possible to operate the four engines on each shaft as one. The operating panel for all engines was found at the back of the engine room. The controls could be divided in controls for each of the engines and the controls needed if this engine was coupled. Hydraulic remote controls consisting of levers and indicators were placed together on the manoeuvring stand.

Two engineers operate the engines of mv Willem Ruys *from one control panel. Earlier, eight engineers would have been needed for the same actions. The picture of a contemporary advertisement from the engineering company that was responsible for the lay out of the panel shows an engineer posing at the right hand side of the panel. The left lever is unmanned. The lever has five positions. A. Vertical is stop B. stand by, ahead and astern and C. power ahead and astern. Beside the name Sulzer in this advertisement a picture of the 8x8 engine blocks as a transparency, seen from above and a rougher sketch than the picture on page 20.*

Control room *(picture M.C.Bakker)*

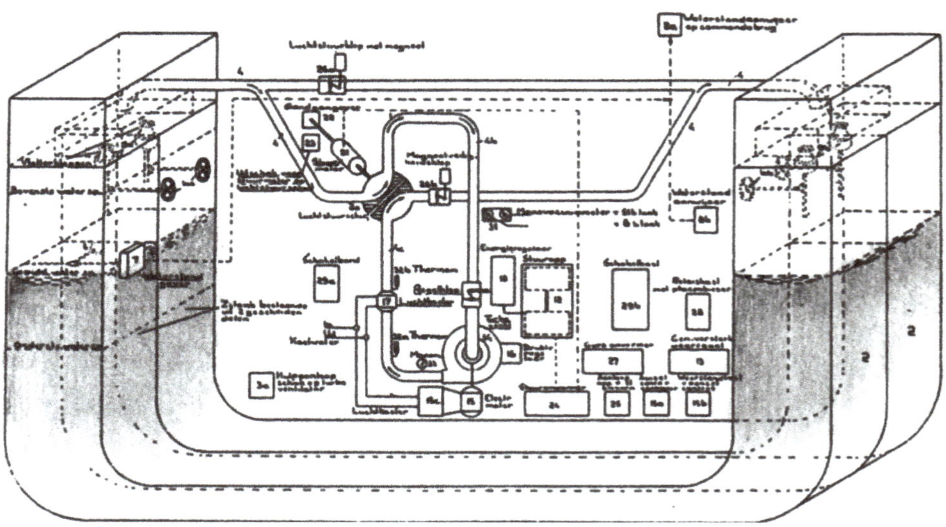

mv Willem Ruys *was the first passenger ship with the system of activated roll tanks. These are based on the principle that the water in the tanks moves in such a way that it counterbalances the rolling movement of the ship caused by the waves. An ingenious installation pumps large quantities of water by means of compressed air in the tanks deep in the ship.*

For the propulsion to be 32.000 shaft horsepower three propellers are needed. The extra cost proved to be prohibitive so that two propellers would have to be sufficient. However that would mean 16.000 hp for each propeller. A capacity that could not be achieved in a single controllable engine. The search for the most economical and powerful propulsion led to the division over a number of cooperating engines that service the propellers indirectly.

Propeller foundry Lips in Drunen is a byword in the shipping world. The company cast the largest propellers ever manufactured in the Netherlands for mv Willem Ruys. *They each weighed 30 tons.*

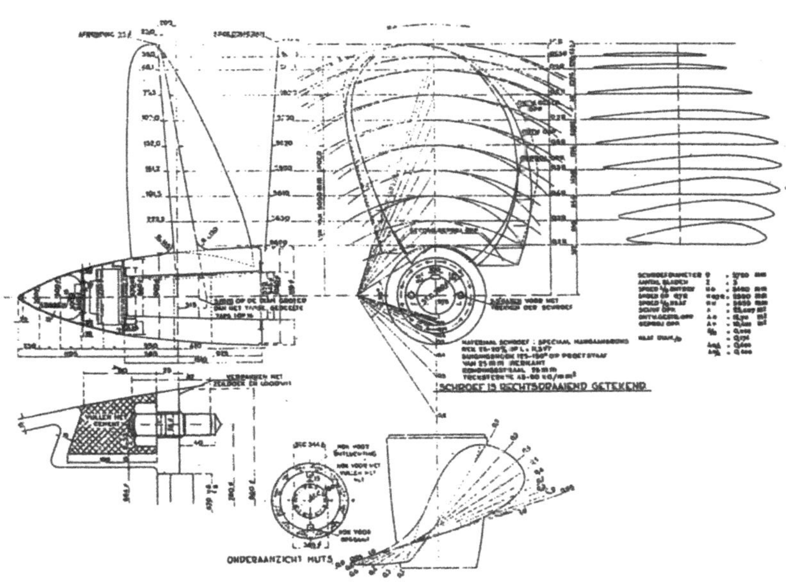

Technical data

Willem Ruys

Tonnage	21.119 gross 12.010 nett 4.925 deadweight
Measurements LxWxB	192.38 x 25.01 x 12.87m draft 8.96m
Engine	8x8 cyl. 2ew Schelde-Sulzer. 23.000Hp 22knots
Passengers	334 I, 320 II, 131 III 74 IV total 869.
Ordered	16 August 1938
Keel laid	25 January 1939
Launch	1 July 1946
Trials	1 - 2 October 1947
Delivery	21 November 1947; total cost: Dutch Guilders 20.000.000
Conversion	15 Nov -5 Feb 1959. 23.112 gross-12401 nett
Cost conversion	Dutch guilders 10.000.000
Passengers	250 I, 850 II
Distance sailed	3.509.000 kilometres.
Total no. of passengers	175.000
Sold to Italy	6 January 1965. Sales price Dutch guilders 33.250.000
Rechristened	*Achille Lauro*
Fate	30 November 1994 on fire. 2 December 1994, sunk.

M.S. Willem Ruys

The Dutch merchant fleet counted 466 ships over and 437 ships under 500 tonnes gross. In total 903 ships when *Willem Ruys* entered service.. A total of 2.385.378 tonnes gross. Eight years earlier before World War II the figures were 1129 ships totalling 2.853.961 tonnes gross. Of the Rotterdam Lloyd fleet twelve out of the thirty one ships survived the war. Rotterdam Lloyd was one of the heaviest hit companies. Rebuilding the fleet was absolutely necessary.

Some more data

Willem Ruys

Oil consumption.	5.500 kg/hr
Measurements main funnel	10.5x5.15x8.4 metres
Length corridors	1.5 kilometres
Surface area promenades	4.200 square metres.
All stairs	1.400 steps
Stoves	713
Sweet water production	300.000 litres
Content swimming pool	60 tonnes
Linen	100.000 pieces
Blankets (wool)	3.500
Crockery	12.000 plates. 10.000 cups and saucers. 20.00 glasses.
Cutlery	10.000 spoons, 7.500 forks. 5.000 knives.
Books	3000
Clocks (electric)	47. all connected to one mother clock.
Elevators	4
Escalators	2
Service bells	870
Fire alarms	67
Alarm bells	55

ROUTE

Dutch passenger liners on the regular service to and from the Dutch Indies sailed from Rotterdam via Southampton, Genua/ Marseille (boat train passengers), Port Said, Suez, Aden, Colombo or Singapore, Sabang, Tandjong Priok, vice versa.
Mail transport had been an important factor in the development of the Dutch passenger transport on this route. When the contracts between the shipping companies and the government were extended the Dutch government demanded faster ships that were built on Dutch shipyards. In 1937 the airplane obtained the right to transport fast mail. The ships kept a part of mail transport which from then on was called sea mail.

In July 1947 the government accepted a proposal which granted KLM the monopoly (for mail transport) to the exclusion of the shipping companies.
'This decision is the beginning of a dark period for Dutch shipping in general and for the Indies lines in particular. The shipping companies sacrificed the main part of their passenger fleet during the war in service of the motherland and are now not capable to maintain communications. Repairs of the fleet cannot be undertaken at the moment due to the high building cost and the shortage of materials.
mv *Willem Ruys* sailed on her maiden trip in December 1947, 'having been under construction since 1939 and miraculously saved from the war while on the slipway.' Under the flag of the Rotterdam Lloyd in name of the Netherlands'. (De Blauwe Wimpel, August 1947 page 185)

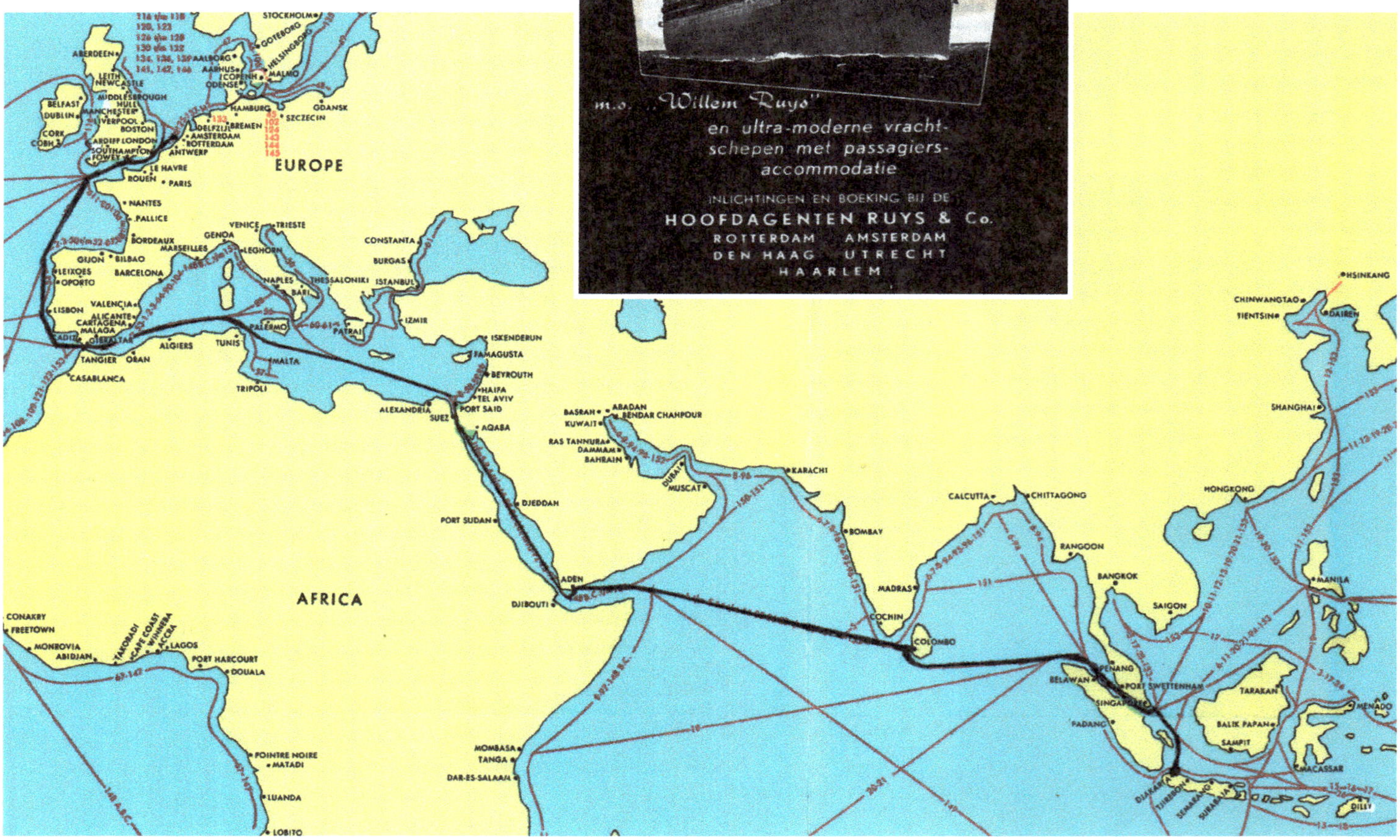

From 1884 the Rotterdamsche Lloyd maintained a regular service to the Dutch - Indies in cooperation with Netherlands Steamship Company (Stoomvaart Maatschappij Nederland). At the beginning of 1940 the Rotterdam Lloyd fleet counted 31 passenger and freight ships totalling 264.911 tonnes gross. *Willem Ruys* sailed in tandem with SMN *Oranje* to and from the Dutch-Indies from 1947 until 1958 always fully booked. The Dutch- Indies became the independent Republic of Indonesia at the end of 1949. This led to increasing tensions between the Netherlands and Indonesia after which the services were cut back. *Willem Ruys* was withdrawn from the Indies route in May 1858 and transferred to holiday cruises. The flagship of the Rotterdam Lloyd would sail the around the world service until some months before she was sold to Italy in January 1958.

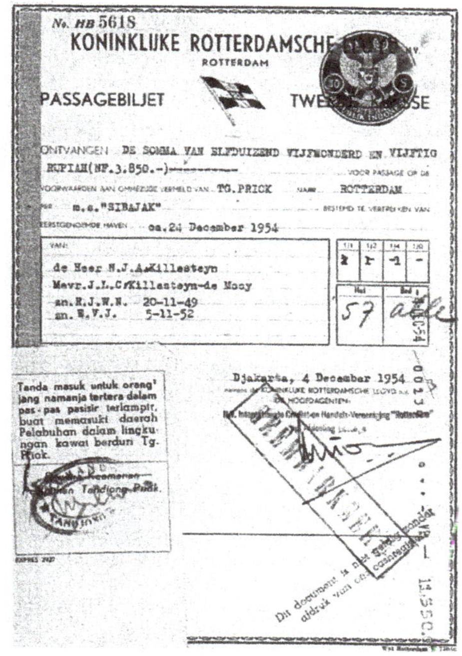
No. HB 5618
KONINKLIJKE ROTTERDAMSCHE
ROTTERDAM
PASSAGEBILJET
ONTVANGEN DE SOMMA VAN ELFDUIZEND VIJFHONDERD EN VIJFTIG RUPIAH(NF.3.850.-)
VOOR PASSAGE OP DE VOORWAARDEN AAN OMMEZIJDE VERMELD VAN TG.PRICK NAAR ROTTERDAM
PER m.s."SIBAJAK" BESTEMD TE VERTREKKEN VAN EERSTGENOEMDE HAVEN ca.24 December 1954
VAN:
de Heer N.J.A.Killesteyn
Mevr.J.L.C.Killesteyn-de Mooy
an.R.J.W.N. 20-11-49
an.E.V.J. 5-11-52
Tanda masuk untuk orang' jang namanja tertera dalam pas-pas pasisir terlampir, buat memasuki daerah Pelabuhan dalam lingkungan kawat berduri Tg. Priok.
Djakarta, 4 December 1954

Rotterdamsche Lloyd ticket for a passage from Tandjong Priok to Rotterdam, departure around 24 December 1954.

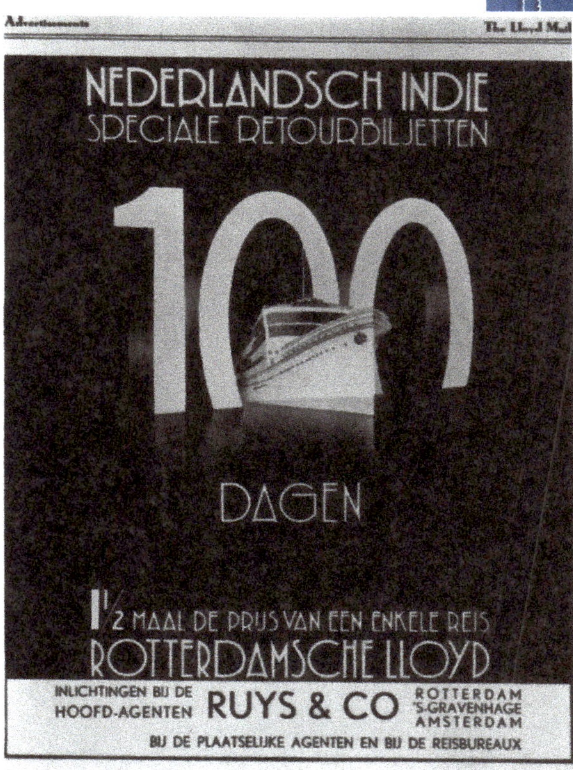

Advertisement in the English Lloyd Mail August 1939. (design. A Guthschmidt). 1938) The advertising department of the company published printed matter that would show the company as favourably as possible. Advertisements, folders and brochures up to press releases and promotional films that would encourage the press to writing illustrated articles about travelling with Rotterdam Lloyd. Pictures of the ships were very important to achieve that goal.

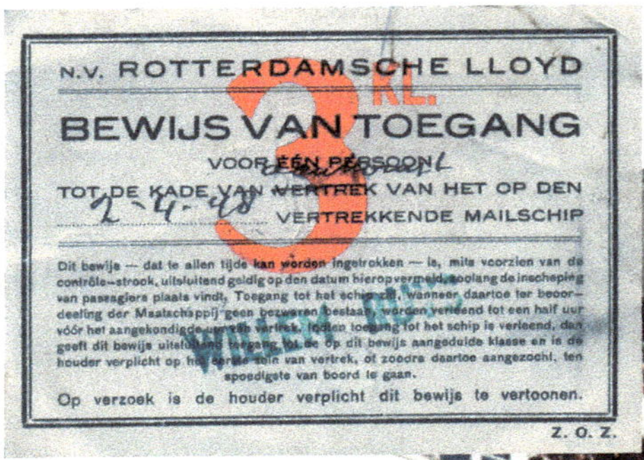

N.V. ROTTERDAMSCHE LLOYD

3 KL.

BEWIJS VAN TOEGANG

VOOR ÉÉN PERSOON

TOT DE KADE VAN VERTREK VAN HET OP DEN

2-4-48 VERTREKKENDE MAILSCHIP

Dit bewijs — dat te allen tijde kan worden ingetrokken — is, mits voorzien van de contrôle-strook, uitsluitend geldig op den datum hierop vermeld, zoolang de inscheping van passagiers plaats vindt. Toegang tot het schip zal, wanneer daartoe ter beoordeeling der Maatschappij geen bezwaren bestaan, worden verleend tot een half uur vóór het aangekondigde uur van vertrek. Indien toegang tot het schip is verleend, dan geeft dit bewijs uitsluitend toegang tot de op dit bewijs aangeduide klasse en is de houder verplicht op het eerste sein van vertrek, of zoodra daartoe aangezocht, ten spoedigste van boord te gaan.

Op verzoek is de houder verplicht dit bewijs te vertoonen.

Z. O. Z.

MAIDEN TRIP

dms *Willem Ruys* sailed for the first time to the Dutch-Indies on December 2 1947. Fully booked carrying 871 passengers. The journey would have its problems. The scarcity of building materials that had taken a great deal of trouble to be obtained meant that these were often of inferior quality. This caused breakdowns. The civil staff had to be broken in, not surprisingly as everybody lacked experience after the long war years. Four days into the passage the anti-roll installation broke down and many passengers fell seasick. The coordination between the four services on board, Nautical, Technical, Civil and Medical needed to be improved.

The passage ran according to schedule. *Willem Ruys* entered Tandjong Priok for the first time on December 24 1947 received with an ear deafening chorus of whistles and ships' horns. The next day the captain C.H. Vellenga entertained about 500 guests, among them the Governor General of the Dutch colony. The captains of the Dutch ships that were lying in Priok had not been invited. Much to their anger and regret. They still navigated in ways reminiscent of the 17th century days of Michiel de Ruyter and the VOC (Dutch East India Company) and they would have loved to be informed about the modern means of navigation that *Willem Ruys* had at its disposal. Such as the two radars, gyro compass, depth sounding, Decca positioning, course writer, rudder alarm, smoke and fire detection including a Sprinkler system.

The radar equipment on the funnel deck had been supplied by Radio Holland. The operating controls were placed on the bridge.

Willem Ruys departed for home on January 3, 1948. She arrived at the Rotterdam Lloyd quay the 23rd of the same month. Teething problems had surfaced and been solved. That is part of a maiden trip. The passengers thought the ship to be most agreeable outside as well as inside, especially the second class saloon was thought to be most impressive.

In the ten years that followed *Willem Ruys* plied the beaten track between the Netherlands and the Dutch-Indies. In 1955 she had logged 707.732 miles. Four times round the world. She sailed 1.437 days out of 2.500. One year later the distance in miles sailed since 1947 equalled that of five times to the moon. She had then been at sea for 1820 days.

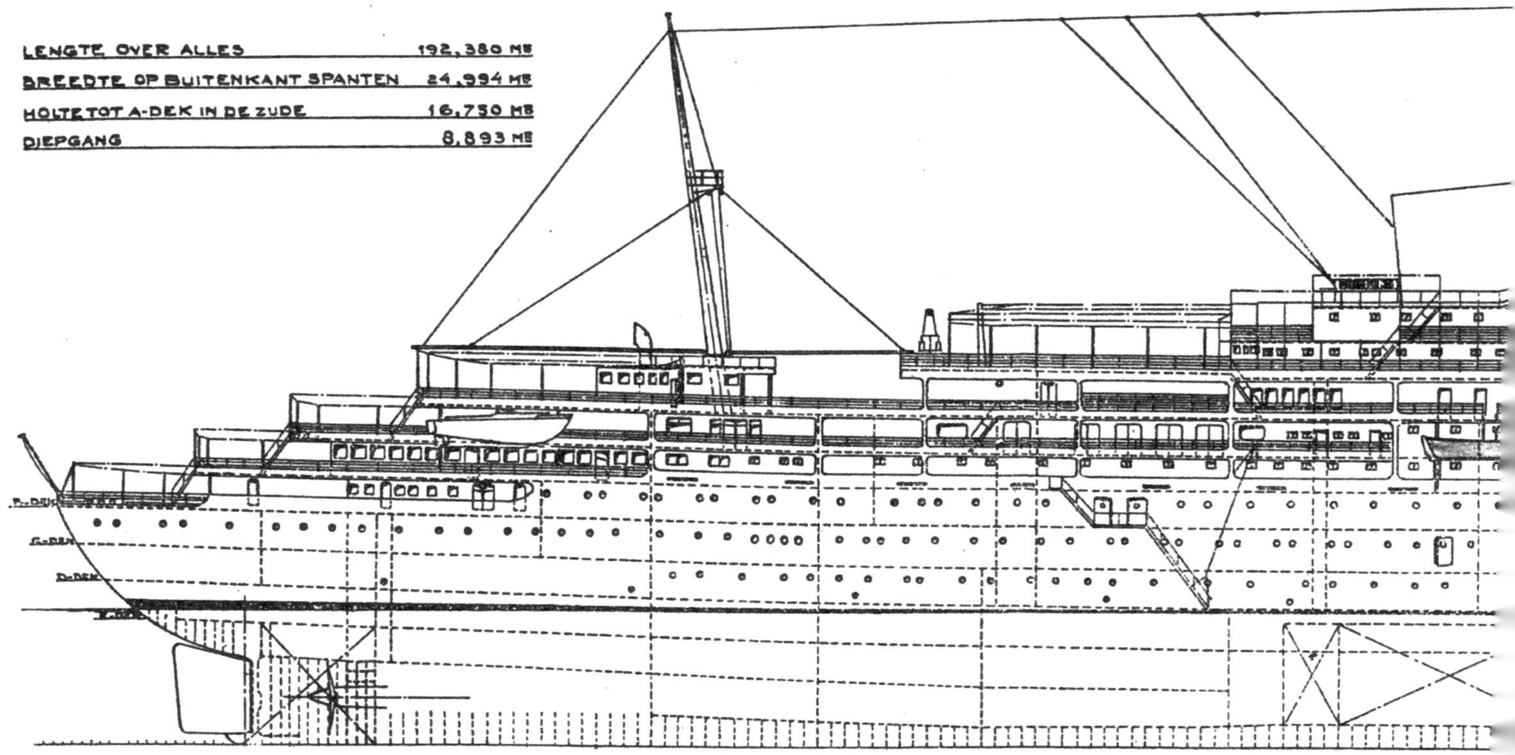

LIFEBOAT ARRANGEMENT.

The lifeboat arrangement deviates from the usual arrangement. The boats are situated half way up the ship's side. There are a number of advantages to this. The collective point of gravity of the boats taken together is low in relation to the metacentric height of the ship. The boats need to be lowered from a height of only 10 metres.

The twelve light metal lifeboats (9,15 x 3,33 x 1,37 m) each have a capacity of 99 persons. They are placed in two rows of six. They are Fleming patent lifeboats. This means that they have a propeller which is powered by arm movements of the passengers. There is no need to row. Furthermore there is a 48 person motor lifeboat and a luxury motorboat for 36 persons. The point of assembly for passengers in case of emergency is the wide promenade deck.

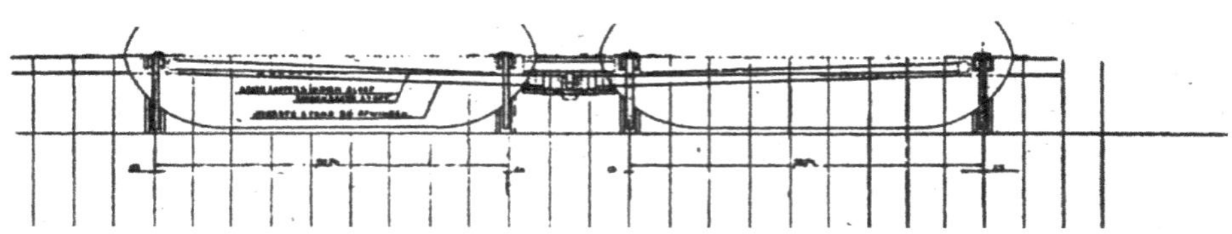

Position of the lifeboats seen from the top and the side.

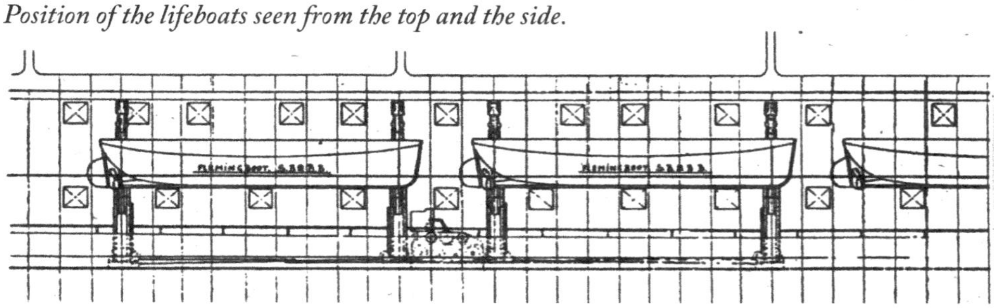

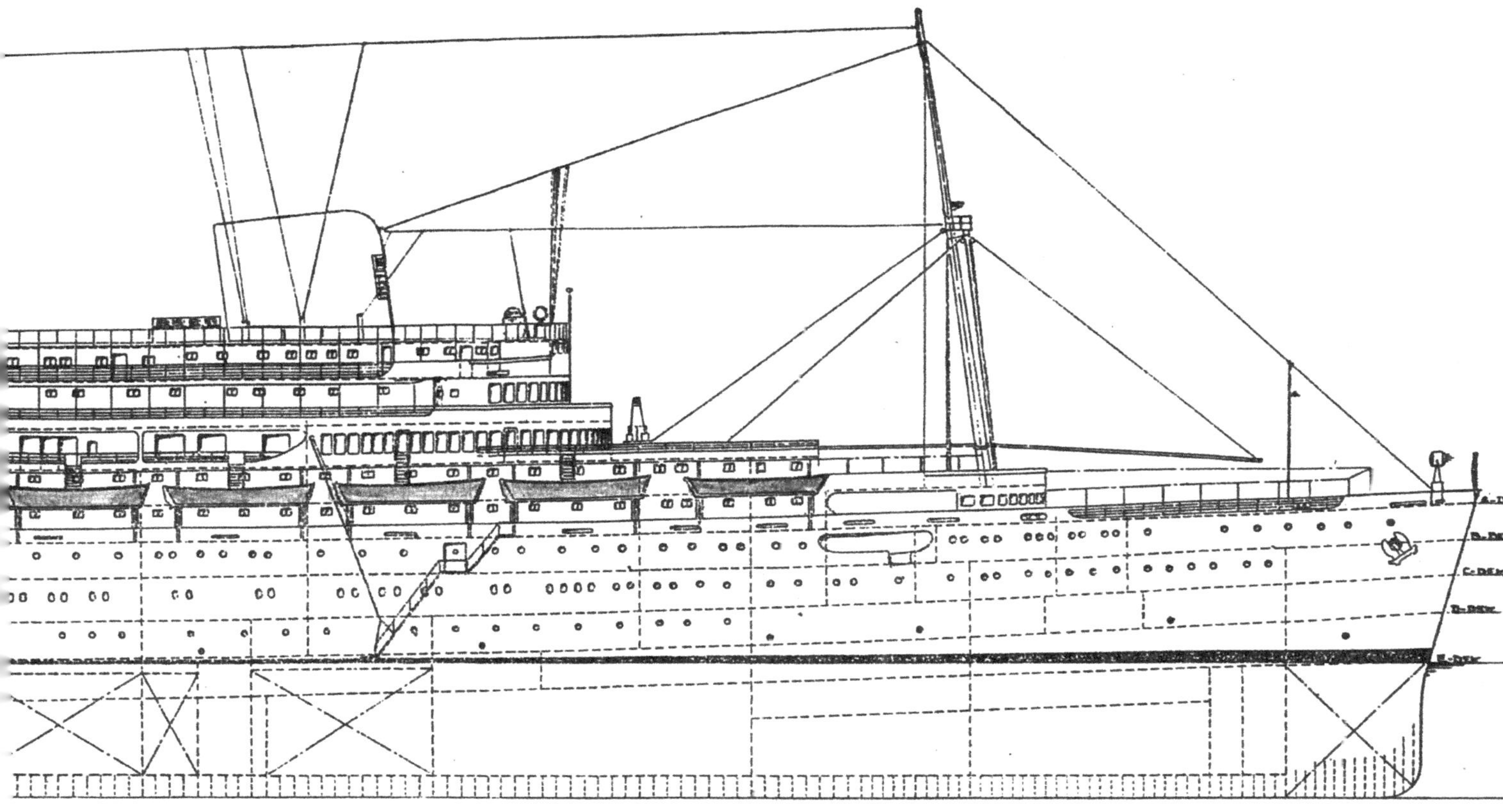

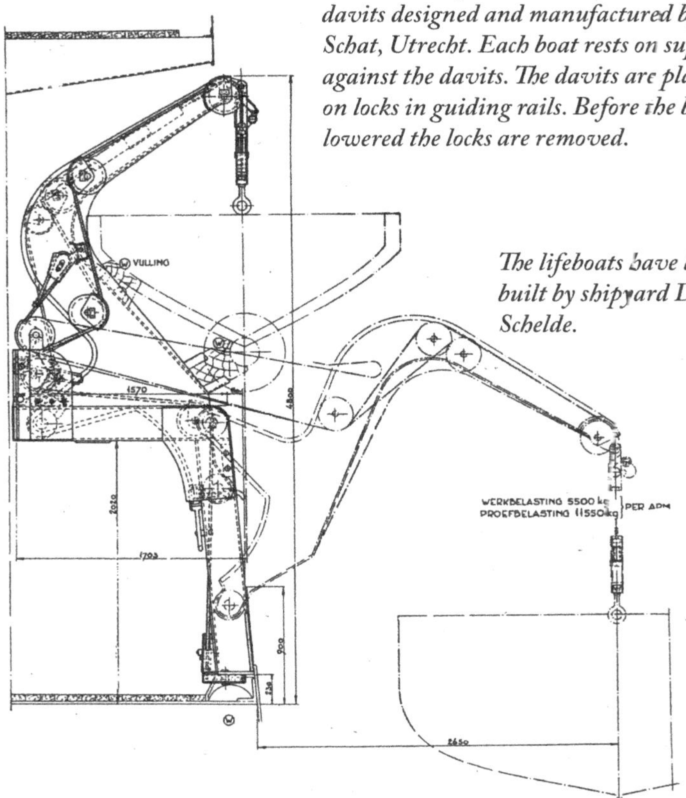

The boats are placed under quadrantal davits designed and manufactured by Co. Schat, Utrecht. Each boat rests on supports against the davits. The davits are placed on locks in guiding rails. Before the boat is lowered the locks are removed.

The lifeboats have been built by shipyard De Schelde.

DECK

Willem Ruys, *then* Achille Lauro *in the Suez Canal. Bridge wing on starboard.*
(picture A. Zuidhoek)

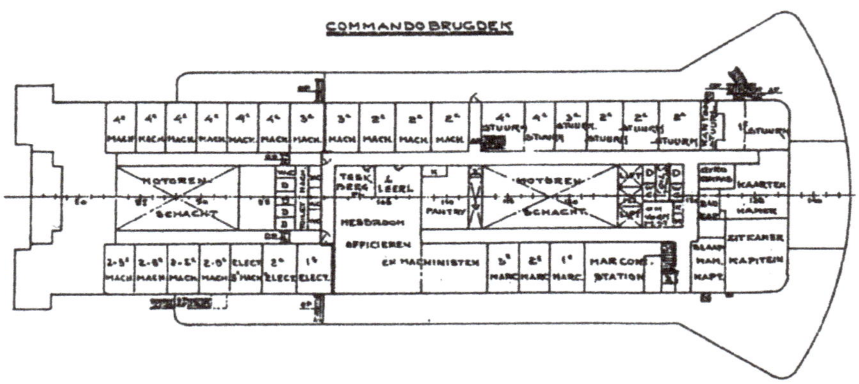

The command bridge deck naturally offers an excellent view to the front, the foredeck and beyond.

Front bulkhead Willem Ruys, *1947. From top to bottom. Funnel deck, command bridge deck, upper promenade deck, promenade deck, bridge deck. Compare with cross section on pages 38-39. The bridge is not situated on the bridge deck (in fact the seventh deck from the bottom). It has three luxury cabins, smoking and social hall second class and the lifeboats.*

The starboard anchoring engine, windlass without the covers. Each anchoring machine can be powered by the engine of the other one by means of a clutch.

The total height of dmv Willem Ruys *from the keel up to the command bridge deck is approx. 27.5 metres. Draught is 9 metres so the total freeboard is approx. 18.5 metres.*

The command bridge deck has besides the wheelhouse and the chart-room the captain's apartments, the cabins of the deck officers, the wireless operators cabins and their wireless hut.

Two vertical anchor winches that also serve as capstan. Each windlass is powered by an eighty HP engine. The driving gear and electrical parts are situated under the deck.

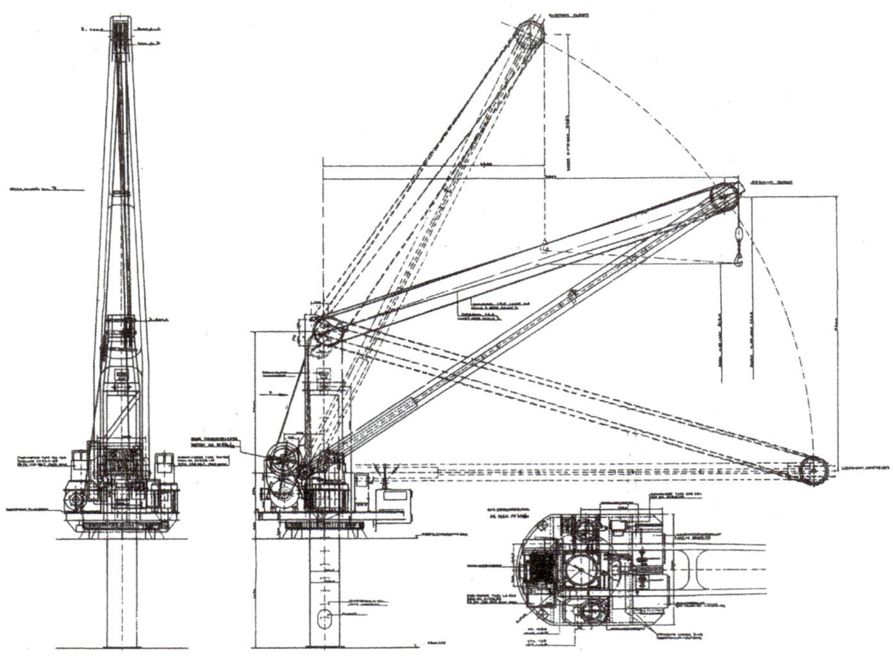

Willem Ruys *has four cranes, two with a lifting capacity of two tonnes at the back and two with a capacity of three tonnes on the foredeck just in front of the front bulkhead. The maximum flight is 9 metres, minimum flight 4.5 metres. The hoisting engine is equipped with a centrifugal break and a built in electromagnetic plate brake. The 17 hp engines, machines and winches have been placed as low as possible on the swing crane platform. The tropicalized electric installation is from Smit Slikkerveer, manufactured by the Nederlandsche Staalindustrie in Rotterdam.*

INTERIOR

Willem Ruys is the last ship with an unmistakable ship like character. Also in the decoration of the interiors. Doors, stairs, wainscoting, partitioning, bulkheads everything was effected by the sheer lengthwise and the rounded decks. The architect could rarely go by a straight line. *Willem Ruys* escaped in 1947 just in time from the later 'straight as she goes' lay out in which the specific aspects of a sea going vessel were sacrificed to those of a land based building. The interiors of *Willem Ruys* had been designed and executed with a skill that did not look for the extravagant or pompous but in a stylish, restrained combination of lines and colour. Interior architects involvesd were F.A. Eschhauzier, H.P. Mutter (sr and jr) and M.J. Luthman.

Movable air- conditioning.

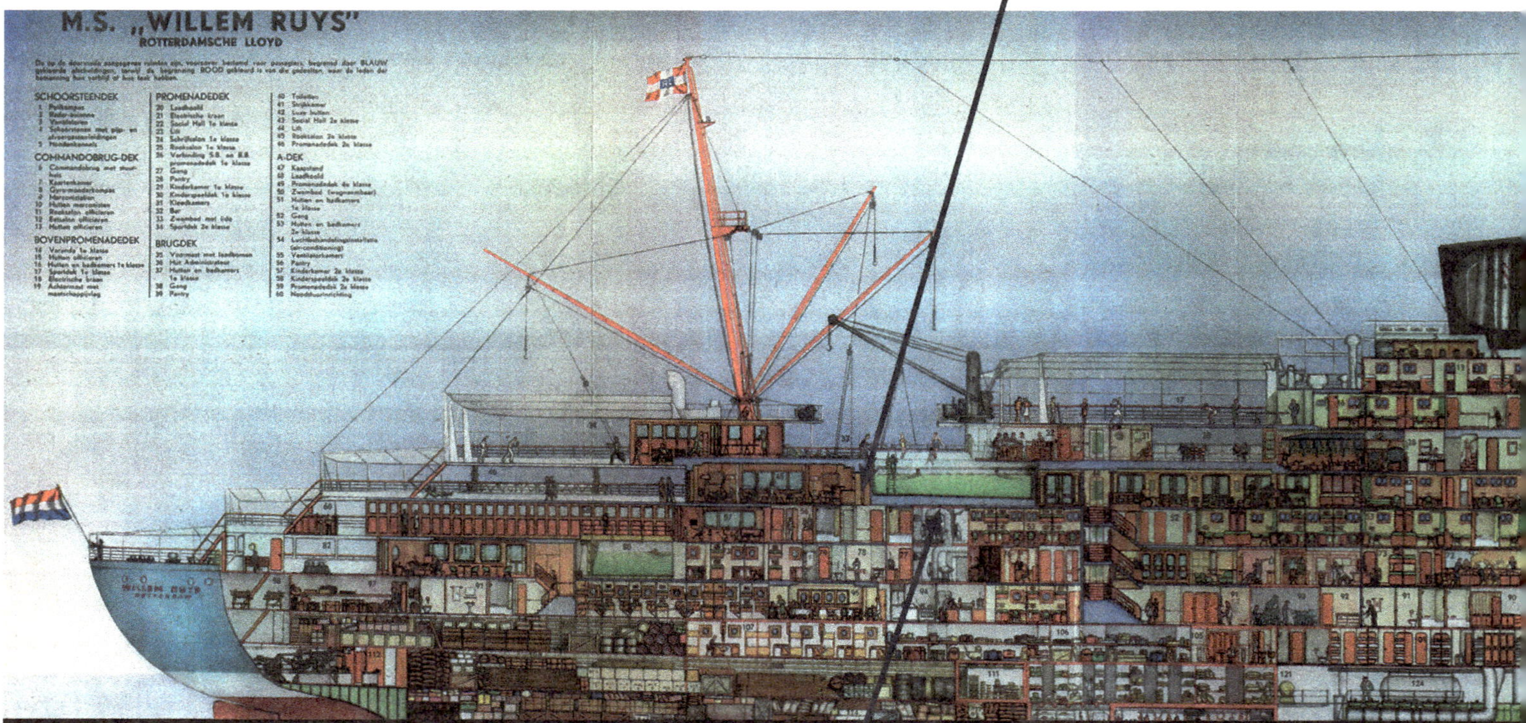

Pool (design: M.J. Luthman) on the promenade deck.

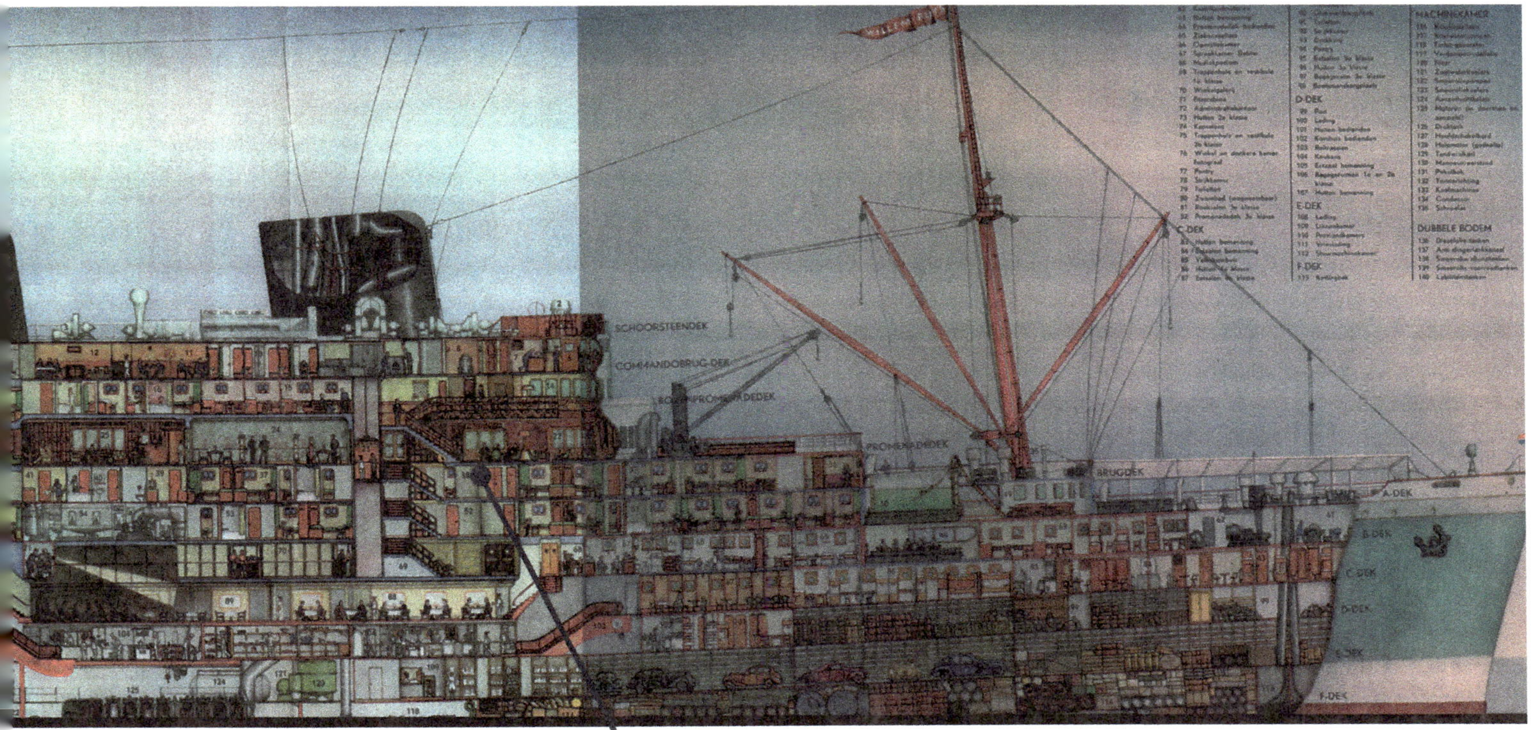

The wainscoting, embellishments and furniture had been designed and executed by the carpentry workshop of the yard. The partitioning of the smoking hall was filled with panels of Italian walnut, the monumental staircase was a masterpiece of furniture making art. *Willem Ruys* boasted rosewood and bronze banisters, enriched with cut motives in gold lacquer. The library doors ornamented with copper braiding. The whole interior was outfitted with decorative windows, statues, tapestry and stained glass panels.

Social Hall first class. After a design of F.A. Eschauzier. The sirens on either side of the stairs were cut by T. Leeser.

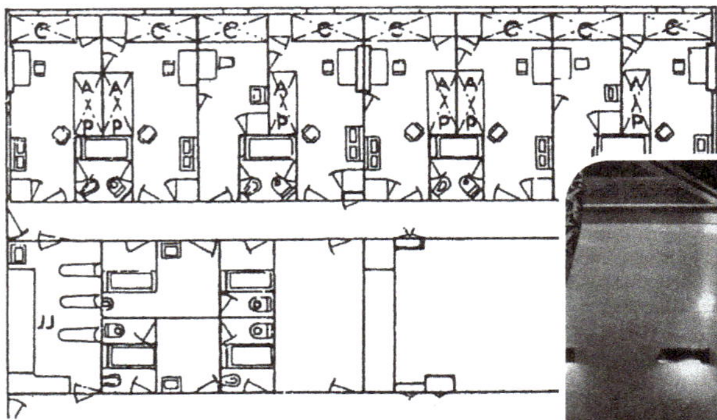

Luxury Cabin on Port side after the design of H.P. Mutter.

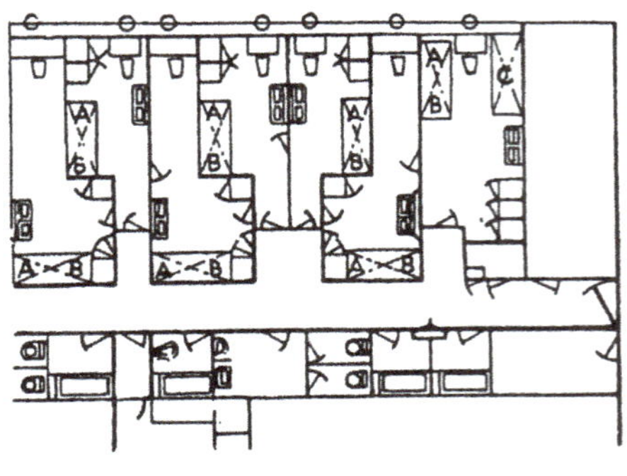

Second class cabins.

Shaft to the hold in front of the first cranes. *(picture A. Zuidhoek.)*

The Rotterdam flagship showed off the latest in technology. Such as electric escalators, automatic doors, dishwashers in the galleys. Each cabin deck had an ironing room. The uppermost deck had a dog kennel and the animals could be walked there.

Willem Ruys *was one of the last Dutch passenger ships designed and built according to the rules of the elements incoporating architecture. Above the flowing lines of a promenade deck (starport) to the stern.*

Below, left: *The rounded corners on the same deck backwards.*

Below right: *The beautifully finished windows and doors to a promenade deck emphasize the urge for good taste in the post-war years.*

(Photo: author)

Assisted by tugs, Willem Ruys *is heading for a mooring spot at the Rotterdam Loyd quay in 1947.*

Emergency steering stand. *(Photo author)*

The wings at the stern (this picture shows starport) (see also pag. 43) formed one of the surprising details of the design. The landings had to go with the conversion to Achille Lauro.

'Monday afternoon's a la Carte lunch offered a sauerkraut dish that was on all the guests' lips.

ATMOSPHERE.

The Head stewards, stewards and the Maitre d'Hotel were Dutch. They managed about 200 staff plus 35 cooks also Dutch. The cabin boys were oriental. The men in the laundry department were usually Chinese. Service on deck and in the dining halls was carried out by Djongos, Indonesians from the island of Madoera, North of Java. Their manager was called 'Mandoer"(boss). After 1958 the Indonesians left the ship.

Cooperation between the different nationalities was good. The service on dms *Willem Ruys* was as on nearly all Dutch passenger liners internationally highly praised. Especially British people chose to travel on Dutch liners. Besides the service they especially praised the efforts of the cooks.
At the Christmas Dinner the dining halls of all the classes were open to each other and all passengers were served at the same time.

Djongos in the galley queueing up for the food to be served.

The atmosphere on board was usually one of an agreeable quietness. The absence of land with all its bother saw to relaxation. The Dutch passengers who were on their way to the Indies could quietly prepare themselves for the tropical heat and the "certainty that useful work was waiting for them"
If homeward bound for a period of leave they were looking forward to seeing their fatherland.
However those who were forcibly repatriated especially in the years after Indonesian Independence often felt depressed or insecure.

Compared to nowadays a ship like *Willem Ruys* did not offer much in the way of entertainment. One could play deck tennis, shuffleboard or quoits. Many passengers brought decks of cards, bingo or board games. Professional artists were not engaged until after 1960. Before that passengers had to make do with a very modest orchestra of about four musicians. A "sportsmaster" conducted a program of collective gymnastics, boxing and swimming. All through the day "nurses" organized activities for the children.

The design of the dining hall is restrained and tasteful.

In the evenings one could dance. Evening attire required.

„WILLEM RUYS"

JANUARI 1955

hier werd zowel
concert-amusement
als dansmuziek verzorgd.
Ook met het
Cabaret-trio kwam het
tot prettige en succesvolle
samenwerking,
waaruit gastvoorstellingen
in Indonesia en Singapore
voortvloeiden,
hetgeende
goodwill der
maatschappij
versterkte.

Orkest
JOHAN PAAY
&
SOLISTEN

KONINKLYKE ROTTERDAMSCHE LLOYD

A page from bandmaster J.Paay's scrapbook.

A British passenger who once travelled to Batavia and back for pleasure said 'now I know that a holiday by car, however lovely it may be is very demanding compared to a journey at sea. What the stressed brains really need is a range of rather uneventful days in which the static and dynamic element are in an equilibrium. So, there is absolutely nothing against me sitting here in the sun for hours doing nothing except watching for flying fish'.

Silhouette Oranje *(top) and* Willem Ruys.

mv Oranje *(1939. 20.016 gross) flagship of the Amsterdam based Stoomvaart Maatschappij Nederland ("Netherlands Steamship Company")* .

COLLISION

It was a tradition for the passenger liners outward and inward bound to meet and pass each other for the pleasure of the passengers. The ships would turn toward each other and then cross at a distance of approximately one nautical mile. A good opportunity to sound the ships' horn. Passengers and crew would wave and cheer. In the night of 5 to 6 January 1953 *Willem Ruys* collided with mv *Oranje* the flagship of Netherlands Steamship Company in the Red Sea. Although both were damaged they could continue their journey. In itself a recommendation for the sound Dutch shipbuilding. After arrival in Rotterdam *Willem Ruys* went to the Wilton Feijenoord shipyard for repairs. She sailed again on March 4.

On 1 January 1955 *Willem Ruys* had logged 707.732 nautical miles. 1310.720 kilometres. That is 30 times the circumference of the earth. A year later on the 27th of April 1956 *Willem Ruys* saw on approaching Naples Vesuvius loom ahead. It was her fiftieth trip. In eight years 87.400 passengers had been transported and sailed 1.645.000 kilometres. Five times the distance from the earth to the moon. According to the press service of the Rotterdam Lloyd the passengers had smoked 61.250.000 cigarettes and 567.00 cigars during that period.

CONVERSION

In 1959 both the Rotterdam Lloyd and Mij 'Nederland' ended the passenger services on Indonesia, the former Dutch- Indies, for political reasons. This was the end of the regular passenger and mail service that had been maintained between Europe and the Far East by the Lloyd since 1875. *Willem Ruys* had to be adapted for a new task.

Two pictures of dms *Willem Ruys*. The top one before and the bottom one after the conversion. (15 September 1958 - 27 February 1959) The top picture shows *Willem Ruys* leaving Tandjong Priok on 3 January 1948. On the bottom picture she is manoeuvring away from the Lloyd Quay, Rotterdam 7 March 1959.

Both funnels had been heightened. 3,5 and 3 metres respectively. The wide white piping had disappeared and her slim profile was to a certain extent lost because some of the decks had been filled up with 107 new cabins. The wings on the stern remained.

The new profile of Dms Willem Ruys in 1959. (drawing. D. Borchert.)

1959. The 'new' Willem Ruys, *portrayed by V.Trip.*

m.s. „WILLEM RUYS"

Gezagvoerder : H. van den Heuvel

ROND DE WERELD REIS

van 16 maart tot 19 mei 1964

(Afstanden in zeemijlen)

1 zeemijl = 1852 meter

★

m.v. "WILLEM RUYS"

Master : H. van den Heuvel

ROUND THE WORLD VOYAGE

from March 16th till May 19th 1964

(Distances in nautical miles)

1 nautical mile = 1852 metres = 2026 yards

MELBOURNE – SYDNEY	590
SYDNEY – WELLINGTON	1250
WELLINGTON – BALBOA	6603
BALBOA – CRISTOBAL (Panama Kanaal)	41
CRISTOBAL – PORT EVERGLADES	1276
PORT EVERGLADES – BERMUDA	949
BERMUDA – SOUTHAMPTON	3025
SOUTHAMPTON – ROTTERDAM	259
TOTAAL/TOTAL	13993
ROTTERDAM – MELBOURNE	12576

'Round the world' cruise leaflet.

Willem Ruys was improved to a value of 22.000.000 guilders. A 15 kilometres air conditioning system throughout the ship. Stabilizers. A new pool with lido decks. A theatre. 107 new cabins. Four new auxiliary engines. Two fully automatic elevators. Etc. The renovations changed tonnage from 21.119 to 23.112 gross. Dutch Maritime magazine De Blauwe Wimpel' was of the opinion 'We need not have any doubt that they (both company and sailors) will succeed in obtaining an important position in the round the world travel market'. Because that was what the Lloyd management had decided. Round the world via Australia and New Zealand. A route with a lot of British competition.

On 7 March 1959 *Willem Ruys* left for its first trip round the world in cooperation with Stoomvaart Maatschappij Nederland ("Netherlands Steamship Company"). 1004 passengers had booked for this trip. *Willem Ruys* sailed East.
Her colleague *Oranje* took a route West Netherlands Steamship Company *Johan van Oldenbarnevelt* sailed East again. After Suez *Willem Ruys* docked at Colombo and Singapore as formerly. From there she sailed to Melbourne, Sydney and Wellington, across the Pacific via Callao. Sometimes to Panama. Then Miami, Bermuda, Southampton and Rotterdam. Duration 63 days.

dms. Willem Ruys *in the Panama Canal. 1959.*

SOLD

The around the world service proved to be profitable at first but was frustrated by the rising cost of fuel, food and the rising crew fees. After some years passenger travel proved not to be possible any more under the Dutch flag for the Rotterdam Lloyd. Already in 1963 *Willem Ruys* lend a helping hand to foreign shipping lines like P&O and the Europe-Canada Lines. And she did some holiday cruises. A year later she was pulled out of service and sold.
'Men do not cry' one of her crew wrote on 22 November 1964, 'but many men who sailed her would have trouble to fight back their tears when they heard that she was going to leave us'.

On January 1965 the Neapolitan ship-owner A. Lauro bought *Willem Ruys* from the Rotterdam Lloyd for 33.25 million guilders (15.09 million Euro). The new owner was so proud that he had the new name painted temporarily on the bows and the stern the same day *Achille Lauro*.
Two days later the former Indies liner left Rotterdam direct for Palermo Sicily to take up her new role in the Italian merchant fleet.

A-deck stern *(picture A. Zuidhoek.)*

Willem Ruys *would be drastically modified to meet the Italian taste. The shaft to the first cargo hold remained in place. Anything in front or behind would change beyond recognition.*

'A good ship, she lay in the water beautifully'

'She is part of my life'.

'It is better like this. The Ruys got a dignified sailor's grave. She lies on the route that she followed for so many years'.

Willem Ruys has always been a lucky ship. A colossal building thatt stood on the slipway for years in a period that it rained bombs all around her. That is being born under a lucky star.
As a Dutch ship she made 62 round trips to the Indies, 2 to Canada, 26 round the world trips and 17 tourist cruises. In total she logged 3.509.000 kilometres and transported 175.000 passengers.

After the sale in 1965 she sailed without fail for the Laura organisation. She led an adventurous life until being over 47 ! years old, not counting the years on the slipway. She caught fire and sank close to her old beaten track between Socotra and Ceylon on 2 December 1994.

References

Websites:
- *www.krlmuseum.nl*
- *www.maritiemmuseum.nl*
- *www.willem-ruys.nl*

literature on dms Willem Ruys:
- *De Blauwe Wimpel, Dutch Maritime magazine year 1947*
- *Een schip vaart uit. Rotterdam 1947 (including cutaway and building plans)*
- *F. Kruidinga – Willem Ruys. Amsterdam 2007*
- *P. Quite – Willem Ruys. Vlissingen 1992.*
- *Ver. van Technici op Scheepvaartgebied - M.S. "Willem Ruys". Rotterdam 1947 (cross sections and, bouwtekeningen)*

Literature algemeen:
- *E.P. de Groot – Varen op de Oost. Alkmaar 1994.*

Translation
Hans Lok

Graphic design
Jantinus Mulder

Author
Arne Zuidhoek

Publisher
Lanasta

First print, September 2014
ISBN 978-90-8616-251-2
e-ISBN 978-94-6456-409-9
NUR 465
e-ISBN: 9781003723820
Contact Liners:
info@walburgpers.nl

Lanasta

www.walburgpers.nl/lanasta

dms Willem Ruys. *(gouache: A. Zuidhoek)*

Willem Ruys, *disguised als* Achille Lauro, *on Durban roads, South Africa , 28 February 1994. (picture: A. Zuidhoek)*